Plumbing Services Series

Water

WATER

Plumbing Services Series

3rd Edition

Rob Kavanagh, Peter Miles

R.J. Puffett, L.J. Hossack

National Library of Australia Cataloguing-in-Publication Data

Title:	Water: Mcgraw-hill plumbing services series /Rob Kavanagh...[et al.}
Edition:	3rd ed.
ISBN:	9781743076941 (pbk.)
Subjects:	Plumbing—Textbooks.
Other Authors/Contributors:	Miles, Peter.
Dewey Number:	696.1

Published in Australia by
McGraw-Hill Education (Australia) Pty Ltd
Level 2, 82 Waterloo Road, North Ryde, NSW 2113, Australia
Publisher: Norma Angeloni-Tomaras
Editorial coordinator: Alex Payne
Production editor: Claire Linsdell
Permissions editor: Haidi Bernhardt
Copy editor: Laura Davies
Proofreader: Kathryn Fairfax
Indexer: Olive Grove Indexing
Cover design: Luke Causby, Blue Cork
Internal design: Natalie Bowra, David Rosemeyer and Dominic Giustarini
Typeset in Scala-Regular 9.75/12pt by diacriTech, India
Printed in China on 90 gsm matt art by 1010 Printing International Ltd

Contents

PART 3 WATER SYSTEMS

Preface

This text provides an overview of how we deal with the sources, treatment and control of water. First we look at the common water collection and treatment methods and the processes associated with these, and then explore alternative methods of water supply, such as might be used where a municipal facility is not available.

The text progresses through the entire process, from collecting water from the hydrological cycle to its storage, transmission and distribution, the valves and controls required, hot water treatment, and water supply to rural areas, including a chapter on pumps and pumping. The treatment and collection of rainwater on-site is also discussed and, as with most things, it varies from area to area. Your local manufacturers are a great resource in this respect. They have already done the groundwork regarding authority requirements in order to provide products that meet their needs.

The final chapters deal with insulation and noise transmission with water systems. The basis for the regulatory information in this text has been the standard AS/NZS3500 which is applicable to most areas and adopted by most authorities.

New to this edition

This latest edition is an update of the earlier editions prepared by Bob Puffet and Len Hossack and brings the text into the twenty-first century, although it must be said that many concepts and principles are timeless. As already stated, the basis is AS/NZS3500, and as such, the references to items such as 'locations', 'sizing' or 'testing' have been sourced from this document. Many of the figures from previous editions have been recreated as they still apply to today's requirements, and photographs have been added to further illustrate and support the text.

So what is new? Even though 'it still flows downhill', how it gets there is changing, with new materials such as polyethylene becoming more popular and pumping arrangements becoming more progressive. We researched the latest products and we cover sustainable practices, such as rainwater harvesting, solar heating and embodied energy.

This book is intended to be a guide to supplement your on-the-job experience. Given the limitations on what could be covered, we have repeatedly recommended that you check with your relevant local authorities as to their specific requirements in plumbing and water supply, as these do vary. Enjoy the text!

Rob Kavanagh and Peter Miles

Acknowledgements

The authors would like to thank Andrew Craine for his review of the text, and also Wayne Clayton, Bruce Paulsen and Steve Eckert for their On-site Stories.

In addition, McGraw-Hill would like to thank the following for permission to reproduce their images:

- Water Corporation (Fig. 1.9)
- Ivor Ebdell, Laurie McGing, South Australian Water Corporation (Figs 1.10, 2.2a, 2.11d, 3.15a,b)
- Hydro Tasmania (Fig. 2.2b)
- Goulburn-Murray Water (Fig. 2.3 b)
- Photo by A. Hollingworth © Sydney Catchment Authority (Fig. 2.4a)
- Norma Angeloni Tomaras (Figs 2.5, 2.16, 2.17, 2.18)
- Bidgee (Fig. 2.6)
- Sydney Water (Figs 2.10, 2.11a,b,c, 3.8)
- Nubian Water Systems (Fig. 2.12)
- Designed and drawn by James Atkinson, Aquafiltercorp. (Fig. 2.13a)
- Permeate Partners (Fig. 2.13b)
- Humes (Figs 2.14b, c)
- Global (HOBAS Pipe) Australia, www.globalpipe.com.au (Fig. 2.25b)
- Pipefusion Engineering Pty Ltd (Fig. 2.30b)
- © Lowell Georgia/CORBIS (Fig. 2.30a)
- Rapid Clamps Pty. Ltd (Figs 2.35, 2.36, 2.37)
- © dbvirago - Fotolia.com (Fig. 3.6)
- Sarah Hetherington, Reliance Worldwide A division of GSA Industries (Aust.) Pty Ltd (Figs 4.9, 6.6a,b)
- Craig Wright, Philmac Pty Ltd (Figs 4.10, 4.11)
- Sam Davidson, Methven (Fig 5.24)
- Tim Fisher, Enware Australia Pty Limited (Figs 5.26, 5.27)
- Michael Acraman, Thermotec Australia Pty Ltd (Figs 6.1, 6.2)
- Rheem Australia Pty Ltd (Figs 7.3, 4a,b, 7.8b,c, 7.11a,b,c, 7.12b, 7.13a,b, 7.15, 7.16, 7.17)
- Solar Edwards (Fig. 7.18)
- Peter Giblin, Rothenberger (Fig. 7.20)
- Wendy Robertson, Team Poly (Fig. 8.1)
- Moore Concrete Products Ltd (Fig. 8.7)
- Natasha Ruciack, Grundfos Pumps (Figs 9.16, 9.17, 9.18)

About the authors

Bob Puffett

After serving as the Head of School of Plumbing and Sheetmetal in NSW, Bob Puffett went on to be Director of Staff, Principal and Assistant Director General, TAFE. Bob was made a Member of the Order of Australia (AM) for his contribution to Technical Education as Director of the Sydney Institute of Technology. Following his 'retirement' Bob became National Chairman of Worldskills Australia. He now serves on local community organisations and is a Board member of a NSW plumbing training organisation.

Len Hossack

Len Hossack is a former Head of School of Plumbing, South Australian Department of Technical and Further Education and has been actively involved in his local community and with the plumbing industry in South Australia for many years.

Robert Kavanagh

Robert Kavanagh contributed most of the chapters in this edition of *Water*. Robert currently holds the position of Training Co-ordinator at the Plumbing Industry Association of South Australia. Robert has been a VET plumbing teacher since 1987 and is a qualified Master Plumber with a lifetime of plumbing experience, having entered the industry at the age of 17.

Robert's qualifications include the Advanced Certificate in Plumbing, a Bachelor of Teaching in Adult Education, Certificate IV in Workplace Education and Certificate IV in Training and Assessment. He was responsible for introducing a Certificate I in Plumbing to high schools throughout Adelaide in partnership with the Department of Education and Children's Services (DECS). He is also the author of *Basic Skills* in the Plumbing Services Series.

Apart from his extensive plumbing experience, Robert has also been teaching sailing for many years at North Haven in South Australia. After many years of sailing and study he gained his Master V Certificate in Maritime, which allowed him to gain a position as Second Officer on the sailing ship the *One and All*, sailing in Australian waters.

Peter Miles

Peter Miles contributed chapters 2 and 7 *(apart from Testing and Commissioning of Hot Water Systems, which was written by Robert Kavanagh)*. Peter is currently Head Teacher of Plumbing at North Sydney TAFE and has been teaching since 1988. He entered the plumbing industry at the age of 15, as the trend at the time for those wanting a trade was to leave in Year 10 and enter the workforce, and later moved into the teaching profession after being inspired and encouraged by several of his teachers.

After qualifying as a Licensed Plumber by completing his trade and post trade qualifications at Gosford TAFE, Peter was awarded the Kembla Scholarship in Adelaide for his efforts. His educational qualifications include a Diploma of Teaching, Bachelor of Teaching in Adult Education, and Masters of Teaching in Adult Education. He is also the co-author of *Gasfitting* in the Plumbing Services Series.

Peter's interests centre on his family and church, and on surfing, when he can find the time.

E-student/E-instructor

www.mhhe.com/au/plumbing

The Online Learning Centre (OLC) that accompanies this text is an integrated online product that will assist you in getting the most from your course, providing a powerful learning experience beyond the printed page.

The OLC contains premium online resources. Students can access it by registering the code located at the front of this text. Instructors have access to an additional instructor-specific resource area.

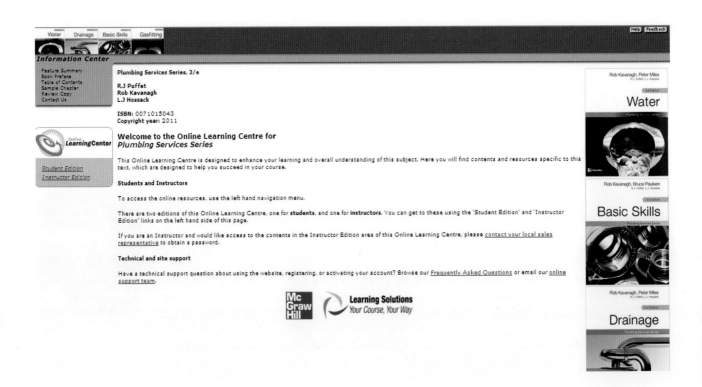

PowerPoint® slides

A set of PowerPoint slides accompanies each chapter and features items that provide a lecture outline, plus key figures and tables from the text.

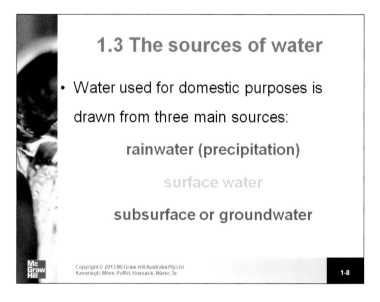

Art Library

All the illustrations from the text are provided in a convenient ready-to-use format.

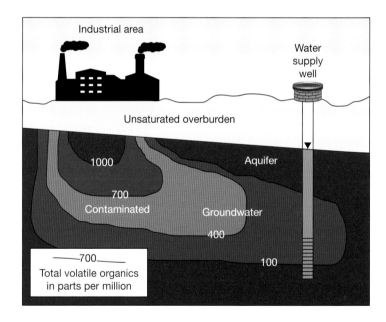

Solutions Manual

The solutions manual contains worked solutions to the chapter exercises provided at the back of the book.

LEARNING OBJECTIVES

LEARNING OBJECTIVES

1 The characteristics of water

PART 1
INTRODUCTION TO WATER

Plumbing Services Series

CHAPTER ONE

The characteristics of water

LEARNING OBJECTIVES

In this chapter you will learn about:

1.1 **the formation of water**

1.2 **the hydrological cycle**

1.3 **the sources of water**

1.4 **the characteristics of water.**

INTRODUCTION

Because of Australia's generally low, erratic and unpredictable rainfall, the amount of water held in storage is of vital importance. This places a considerable burden of responsibility on the water supply authorities and the general public to ensure that limited water resources are neither wasted nor polluted. Plumbers also have a responsibility to ensure that water resources are protected.

Australia's supply of fresh water is becoming increasingly vulnerable to droughts and floods, possibly as a result of climate change, and certainly in response to population growth in our cities (Figure 1.1). As a result, there is a growing emphasis on water conservation, and various regions now impose restrictions on the use of water.

FORMATION OF WATER

Millions of years ago Earth was a ball of white-hot gases, travelling through space. Gradually two gases, hydrogen and oxygen, combined to produce an entirely new substance: water. They combined in a ratio of 2 parts hydrogen to 1 part oxygen—hence the chemical symbol for water: H_2O. Because of the intense heat, the water was produced as steam or water vapour and remained in that state until the planet cooled and the temperature of the surrounding atmosphere had lowered. When the dewpoint of the water vapour was reached, it condensed to form a liquid and fell as rain, filling depressions in Earth's crust. This formed the seas and oceans that currently occupy approximately 70 per cent of the planet's surface.

Water for life

Life developed in the seas and on land, dependent upon water for its existence. No known form of life can exist without water. Life in the sea is continually surrounded by water and obtains sufficient water by a process known as osmosis or diffusion. Land dwellers, however, need a much more complicated and sophisticated system for absorbing and retaining water because they continually lose moisture to the surrounding atmosphere by evaporation and transpiration.

FIG 1.1 Earth requires a regulated supply of water; too much or not enough can have catastrophic effects

All earthbound organisms require a continuous supply of water to replace that lost to the atmosphere.

All of the water currently on Earth—including the contents of the seas, the polar ice caps and the water vapour trapped in the atmosphere—is all that we have; it is, in fact, all that we will ever have. Early humans were under the misapprehension that the rain that fell was a fresh supply. This is not true. It is water that has evaporated from some other place on earth and has been transported through the atmosphere to fall as rain. This transfer of water from one place to another—i.e. water drawn from the sea, deposited on land and then returned to the sea again—is referred to as the water, or hydrological, cycle (Figure 1.2).

FIG 1.2 The hydrological cycle

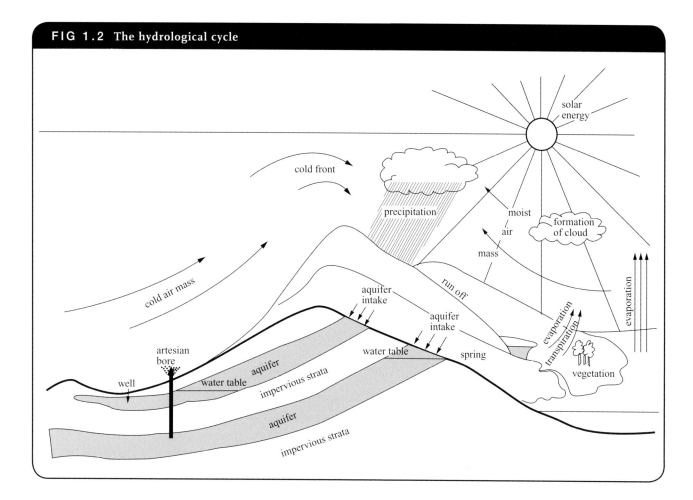

The hydrological cycle

The hydrological cycle is a perfect example of perpetual motion. Although it is often interrupted, it is a continuous process and involves the evaporation, transpiration, condensation and precipitation of water. The operation of the cycle is dependent upon three criteria:

1. The amount of solar energy available. This varies considerably from one geographical location to another. Generally speaking, the amount of solar energy reaching the earth decreases as the distance from the equator increases. However, the type and texture of the land surface and its ability to reflect radiant energy are major factors in the amount and intensity of energy available for evaporation to take place.

2. The ability of the surrounding air to absorb moisture. This factor is of equal importance as it dictates the rate at which evaporation takes place. The amount of moisture that the air can absorb is limited by the temperature of the air. It can also vary considerably from place to place. The measure of the variation is referred to as the 'relative humidity' of the surrounding air. The lower the relative humidity, the greater the amount of evaporation and water absorption that can take place. As the air absorbs moisture, the relative humidity increases and continual absorption requires a drier, less humid air moving in to take its place.

3. The availability of water for evaporation. This availability varies considerably from land to sea. At first glance it appears that seas and lakes would offer the only surfaces from which water may be evaporated. However, the rough surface of land that has been saturated by rainfall will lose moisture by evaporation at an equivalent rate to the surface of the sea or other large expanses of water. This rate of evaporation is short lived. As the surface of the ground dries out, the rate of evaporation and transpiration decreases.

Australia, being an island continent, obtains the majority of its moisture from the sea. Australia is predominantly semi-arid, with its maximum rainfall along the eastern coastal belt. Most of this rainfall returns directly to the sea because of the location of the mountain ranges of the continent. The effect these mountains have on Australia's rainfall can be seen in Figure 1.3.

SOURCES OF WATER

Oceans cover approximately 70 per cent of Earth's surface and contain more than 97 per cent of its water. The remaining three per cent is held in rivers, lakes and streams; below ground as subsurface water trapped in aquifers; as snow and ice where approximately 75 per cent of fresh water is permanently stored; or in suspension in the atmosphere in the form of vapour. As can be seen from these figures, only a small amount of fresh water is readily available.

FIG 1.3 Predominant wind flow over the Australian continent

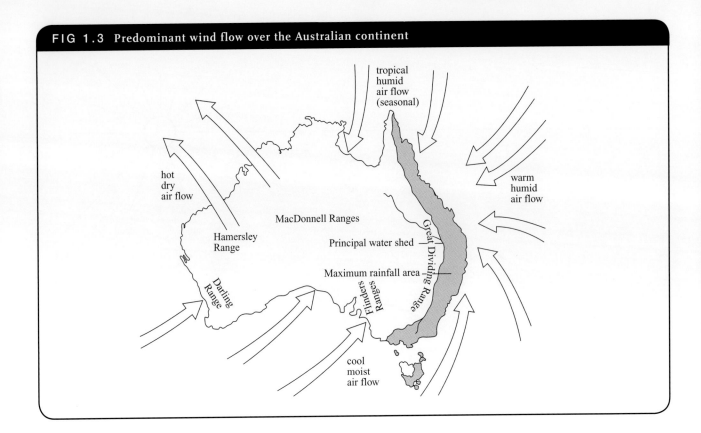

Water used for domestic purposes is drawn from three main sources:

1   rainwater (precipitation)

2   surface water

3   subsurface or groundwater.

Rainwater

When required for domestic purposes, rainwater is generally collected as run-off from buildings and stored in rainwater tanks, either above or below ground.

Surface water

Surface water is drawn directly from creeks, rivers and lakes. Only those creeks and rivers that have a stable flow throughout the year are suitable for water supply via a network of dams and reservoirs.

Surface water can be drawn from creeks and rivers that have an intermittent flow caused by irregular rainfall or annual snow melt. This water is retained in artificial lakes or dams when the flow is at its maximum.

Groundwater

Groundwater, or subsurface water, can be obtained from wells, springs or artesian bores.

Wells

Wells are either driven or dug to below the water table. Water from this source is lifted to the surface by artificial means, usually a pump.

Springs

Springs generally occur where the aquifer or water-bearing stratum appears close to or at the surface.

NOTE: An aquifer is an underground layer of rock through which water is transported and from which water can be drawn.

Artesian bores

Artesian bores are sunk into the aquifer containing the water. Water flows from them under pressure, which varies according to the height of the water table. The water from artesian bores usually contains a high concentration of dissolved minerals that have been leached out of the ground through which the water has passed. Artesian water often discharges at ground level at temperatures of up to 100° C because of the great depths from which it rises, which can be up to 2000 m. The Great Artesian Basin is one of many artesian basins throughout Australia and covers 1 700 000 km²—approximately one-fifth of the Australian land mass (Figure 1.4).

CHARACTERISTICS OF WATER

Before discussing the installation of pipework, storage vessels and equipment used in water supply, it is necessary to have an understanding of some of the characteristics of water and how these characteristics affect the design and installation of water supply equipment. Hydraulics is the study of liquids in motion; hydrostatics is the study of liquids at rest.

FIG 1.4 Major sedimentary basins of Australia

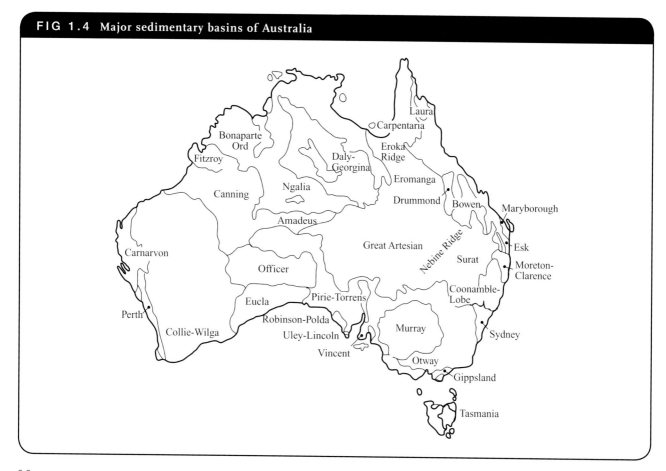

Mass

Water has mass, but this varies according to the water temperature and the chemical composition. For practical purposes it is assumed that 1 L of water has a mass of 1000 g, or 1 kg, at 4° C.

State

Water is obtainable in three states: solid, liquid and gas.

Solid

Pure water freezes at 0° C and changes from a liquid to a solid (ice). During this process it will expand by approximately one-tenth of its volume. This expansion exerts extreme pressures within pipes or storage vessels, causing them to fracture. The amount of expansion can be demonstrated by filling a 25 mm diameter open-topped tube with water, to a height of approximately 100 mm. (The tube allows the expansion to be in one direction only.) Place the tube in a freezer and allow it to freeze. If you then measure the column of water, you will find that it has increased in length from 100 mm to 110 mm.

Liquid

Water in liquid form always adopts the shape of the container that it occupies. In containers that are connected and open to the atmosphere, water will always reach a point of equilibrium, that is, the same level regardless of the shape of the containers (Figure 1.5).

FIG 1.5 Water always reaches its own level regardless of the shape of the container

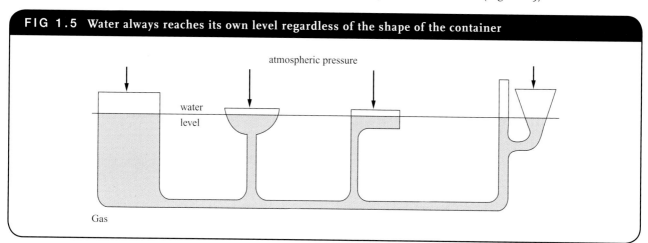

Water also has the ability to dissolve substances. This is called solvency, and the degree of solvency depends on the chemical content of the water.

Gas (vapour)

At sea level the atmospheric pressure is 101.3 kPa. When subjected to this pressure, pure water will boil at 100° C, and during this process it will be converted to steam. As water is converted to steam its volume will increase approximately 1600 times, that is, 1 m³ of water will produce 1600 m³ of steam. If the 1600 m³ steam is allowed to cool below its dewpoint, it will condense and return to water again and occupy 1 m³.

When this occurs in a sealed container, such as a hot-water heater, a vacuum will be created in the space previously filled with steam. The atmospheric pressure which is being applied to the outside of the container would then exceed the internal pressure and could result in the container being crushed if the proper precautions were not taken.

Density

Water is at its maximum density at 4° C; that is, a given quantity of water occupies its smallest space at this temperature. Most materials expand when heated and contract when cooled; however, water expands when heated above 4° C and also expands when cooled below this temperature.

Water has a relative density of 1 and the density of all other substances is determined by comparing a given volume of a substance to the same volume of water at a temperature of 4° C. As an example, if a substance is stated as having a relative density of 0.5, this means that a given volume of this liquid weighs half as much as an equal volume of water at the same temperature.

Pressure

As previously stated, water has mass; however, this is quite distinct from the pressure exerted by water. Pressures exerted on the internal surfaces of a hot water cylinder are far greater than the total mass of the water contained in the cylinder.

The amount of water pressure a gravity-fed water system achieves depends on the amount of water it stores and the height (head) difference between where it stores water and the point where the water is used. An increase in water quantity or head increases the system's water pressure. The pressure exerted by water is dependent on the depth of the water, or head as it is known, above the point at which the pressure reading is taken.

All substances are composed of minute particles called molecules, which are held together by a force known as cohesion. The cohesive force in liquids is very small, allowing liquid molecules to move freely with the forces of gravity. As Earth's gravitational force is downwards, all the water molecules attempt to reach the lowest point of the liquid; as this is not possible they tend to level out in layers. This is the reason why the free surface of a liquid is always horizontal and water adopts an equilibrium

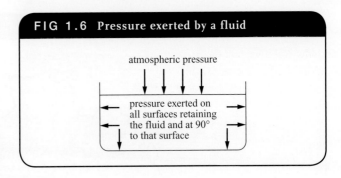

FIG 1.6 Pressure exerted by a fluid

atmospheric pressure

pressure exerted on all surfaces retaining the fluid and at 90° to that surface

(that is, it is the same level regardless of the shape of the container).

The pressure exerted by a fluid is always at 90° to the surface retaining the fluid (Figure 1.6). Pressure exerted on the base of the container is equal to the total weight of the water contained, while the pressure exerted halfway up the container would be equal to half the weight of the fluid contained.

Intensity of pressure

This is stated as being the force created by the weight of a given mass of water bearing on one unit of area, usually taken as one square metre (m²). Therefore,

$$\text{pressure} = \frac{\text{force (weight)}}{\text{area}}.$$

Under the SI system, the unit of pressure is the pascal (Pa). One pascal is equal to one newton per metre squared or 1 Pa = 1 N/m². These basic units, however, are extremely small and it is not practical to use them in plumbing calculations. The more convenient multiples are the kilopascal and kilonewton (kPa and kN), which are 1000 times the basic units.

NOTE: If 1 L of water at 4° C has mass 1000 g (1 kg) and 1 m³ contains 1000 L, then 1 m³ of water has a mass 1000 kg. This mass is subject to gravity, so it has a weight or force of 1000 × 9.81 = 9810 N. Because the water is resting on a surface 1 m², the pressure it exerts is

$$\frac{9810}{1} = 9810 \text{ Pa}.$$

Also, as it is 1 m high, we can say that 1 m head of water exerts pressure of 9810 Pa or 9.81 kPa.

To convert head of water (in m) to pressure (in kPa), multiply by 9.81.

$$\text{pressure (kPa)} = \text{head (m)} \times 9.81$$

Example

Calculate the intensity of pressure exerted by a column of water 20 m high.

$$\begin{aligned}
\text{pressure} &= \text{head} \times 9.81 \\
&= 20 \times 9.81 \\
&= 196.20 \text{ kPa}
\end{aligned}$$

Hydraulic gradient

The flow of water from the open end of a pipe is dependent upon the head of water acting on the outlet and the amount of head (pressure) absorbed by friction created when the water passes through pipes and fittings. Therefore the head pressure needs to be greater than the pressure loss created by the friction. Figure 1.7 indicates the position of the hydraulic mean gradient in a simple system. The gradient is taken as a straight line from the surface of the water to the outlet. Therefore, as the water level falls in the storage tank so the hydraulic mean gradient will fall.

For this reason the gradient line is usually taken from the bottom of the storage tank, to allow a safety margin in design.

In the design of water systems, care must be taken to ensure that all draw-off points are below the hydraulic gradient. If an arrangement as in Figure 1.8 were installed, it is clear that if water was allowed to flow freely from outlet *D*, the outlet at *E* would be starved. However, by correct pipe sizing, that is, reducing the frictional loss in the pipe from *A to E* and reducing the discharge at *D* to the required minimum, water can be made available at point *E*.

FIG 1.7 The position of the hydraulic mean gradient

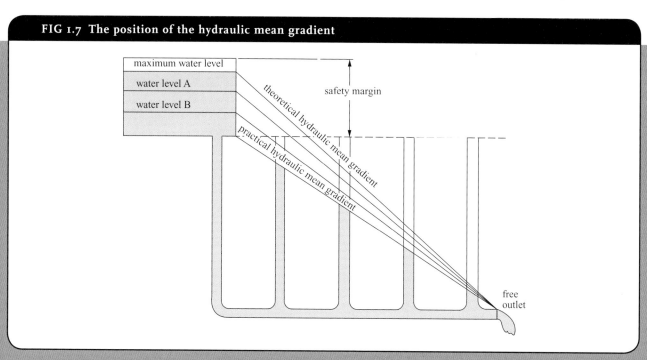

FIG 1.8 Draw-off points must be below the hydraulic gradient

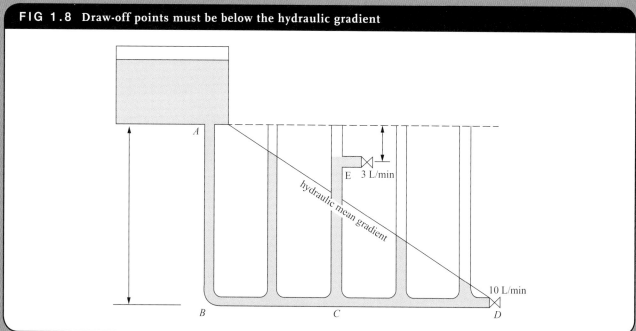

NOTE: If, for example, a discharge of 10 L/min were required at *D* and 3 L/min at *E*, pipe *CD* should be sized so that a discharge of 10 L/min is obtained under a 10 m head, pipe *CE* sized to give 3 L/min under a 2 m head, and pipe *ABC* should be sized to give 13 L/min under a 10 m head.

Compressibility

Water is practically incompressible. This is one of the main causes of water hammer in pipework. If allowed to go unchecked, water hammer can cause considerable damage to pipework appliances and fixtures. It is caused by the sudden arrest of water flowing along a pipe. If a tap is opened, the water flows along the pipe at a considerable velocity. When the water is suddenly stopped, it recoils, as, being incompressible, it is unable to absorb the shock within itself. The pressure inside the pipe is momentarily increased, causing shock waves to travel along the pipework with the energy being absorbed by the pipes and fittings, causing them to vibrate and create noise.

There are many ways of preventing water hammer, such as with the type of control taps used, correct supporting of pipes and the installation of apparatus designed specifically for this purpose. These will be discussed in Chapter 6.

ALTERNATIVE SUSTAINABLE WATER SOURCES

With traditional water resources being depleted, Australian governments and water companies are looking to alternative water sources to meet demand in urban as well as rural areas.

These alternatives include:

- seawater
- saline groundwater
- urban stormwater
- wastewater (e.g. treated sewage effluent).

Seawater desalination

A desalination plant essentially separates saline water into two streams: one with a low concentration of dissolved salts (the freshwater stream) and the other containing the remaining dissolved salts (the concentrate or brine stream). The plant requires energy to operate and can use a number of different technologies for the separation of the saline water.

The most common treatment is reverse osmosis, whereby the salty water is forced through extremely fine membranes. The amount of the feed water discharged to waste in the brine stream varies from 20 to 70 per cent of the feed flow, depending on the technology employed and the salt content of the feed water.

Desalination is becoming more widely accepted as an alternative source of water as the technology improves. Desalination plants can be provided in a wide range of outputs to cater for small isolated communities or to contribute substantially to water supplies for large cities and even for irrigation

In Australia, the majority of distributed water comes from surface water resources. Western Australia was the only state that sourced a significant proportion of its distributed

water supply from groundwater sources. However, this changed rapidly when a large seawater desalination plant was completed in Perth at Kwinana and a second plant became operational in 2012. Many other large desalination plants are being built or considered in other states (Figure 1.9).

FIG 1.9 Kwinana Desalination Plant in Perth

Re-use water

There are a variety of water sources that may be supplied as re-use water, including waste water (from sewerage systems), drainage water, stormwater or other water providers (i.e. a 'bulk' re-use water supply). Sewerage systems collect and treat waste water to primary, secondary or tertiary levels.

Stormwater may also be collected using infrastructure separate from sewerage systems and, depending on its intended use, may or may not be treated before being supplied as re-use water. Drainage water is collected in regional collection drains managed by irrigation/rural water providers. This water may be supplied as re-use water to customers or discharged to the environment. However, in urban systems the water is typically treated before supply (Figure 1.10).

FIG 1.10 Glenelg to Adelaide Parklands Recycled Water Project

FIG 1.11   Aquifer storage and recovery

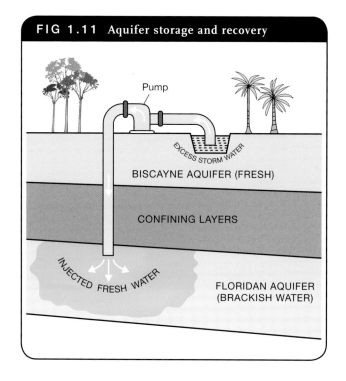

Pump

EXCESS STORM WATER

BISCAYNE AQUIFER (FRESH)

CONFINING LAYERS

INJECTED FRESH WATER

FLORIDAN AQUIFER
(BRACKISH WATER)

Aquifer storage and recovery (ASR)

Aquifer storage and recovery (ASR) is the process of injecting water from Earth's surface into a suitable underground aquifer for storage and re-supply (Figure 1.11). One of the advantages of ASR is that large volumes of water from wet periods (e.g. storm events) can be stored with very little evaporation and can then be used when water is scarce. It is a suitable technology for urban environments where surface storage is limited and demand is seasonal. ASR is likely to become a more common alternative source of water in future.

FOR STUDENT RESEARCH

Research the source/s of the water used for domestic purposes in your own home.

CHAPTER TWO

Water storage, transmission and distribution

LEARNING OBJECTIVES

In this chapter you will learn about:

2.1 the storage of water

2.2 impurities and treatment

2.3 recycled water

2.4 water softening

2.5 transmission and distribution systems

INTRODUCTION

Although it is the water authority's responsibility to provide water of an acceptable quality, it is still important for plumbers to understand the system they are connecting to, and to accept responsibility for maintaining its integrity. Where accredited by the local authority, plumbers may also be involved in the installation of water mains.

Plumbers are also directly involved in systems for recycling water, either by installing them or by connecting to them; in transmission and distribution systems; and in the installation of large pipework for water services. These topics are covered in this chapter.

STORAGE

Water stored for human consumption is usually held in dams or reservoirs built across suitable river valleys. The reservoirs are constructed principally of a concrete or earthfill wall, built to be both watertight and strong enough to withstand the forces exerted by the desired quantity of water. The site chosen for an impounding reservoir depends largely on the following:

- quantity and quality of water available
- elevation of the site
- distance from the consumer
- suitability of the site for dam construction with respect to foundations for the retaining wall
- availability of suitable construction materials
- type of strata and soil composition.

The quantity and quality of water available is of paramount importance. Availability depends primarily on the size of the catchment area where the feed water is sourced, and on the soil type, vegetation and human activities within its boundaries. The quality of the water collected in the catchment area is judged primarily by chemical and bacteriological analyses that ascertain the amount of contamination present. The majority of contaminants or pollutants found in water can be eliminated or modified by some form of chemical treatment.

The elevation of the reservoir above the point of consumption can dramatically reduce the costs incurred in transporting water to the consumer. This is achieved by allowing the water to flow by gravity rather than by pumping it. If, however, a dam or reservoir is constructed a considerable distance from the consumer, it may be necessary to provide pumping equipment to overcome frictional losses within the pipeline system.

Geological surveys are carried out before a dam site is finally selected. This consists of rock core sample testing to determine the extent and type of rock present on the proposed site and its suitability to withstand the enormous pressures that are exerted by the retaining wall and the impounded water. The design of the retaining wall used also depends on the strength of the surrounding geological formation and on the availability of suitable materials, either on-site or accessible from the site. (Figures 2.1, 2.2 (a) and (b), 2.3 (a) and (b), 2.4 (a) and (b).)

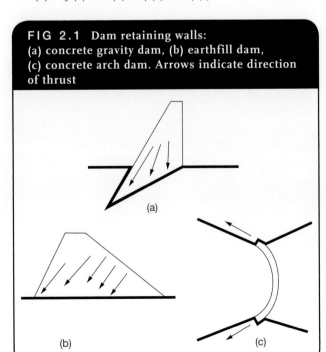

FIG 2.1 Dam retaining walls:
(a) concrete gravity dam, (b) earthfill dam,
(c) concrete arch dam. Arrows indicate direction of thrust

FIG 2.2(a) Mount Bold Reservoir, a concrete arch dam

Mount Bold Reservoir, built on the Onkaparinga River, is the largest reservoir in South Australia. Although not directly connected to the reticulation system, it releases water into Clarendon Weir as required, maintaining an adequate water supply for Happy Valley and parts of the Mount Lofty Ranges. Built in 1938, the wall was raised 6.4 metres in 1962, increasing its capacity to 46 180 megalitres. The height of the wall is 50.6 metres and it is 192 metres in length. The site has been identified as being suitable for a mini hydro-generator.

FIG 2.2(b) Gordon River Dam, a concrete arch dam

The Gordon River Dam is used to generate hydro-electricity in Tasmania. It is the highest concrete arch dam in Australia, the fifth highest dam overall, and is also one of the largest in capacity. It stands 140 metres high and 192 metres long, and is 18 metres thick at its base. The dam has a capacity of 12 450 000 megalitres and supplies up to 13 per cent of Tasmania's electricity.

FIG 2.3(a)   Wivenhoe Dam, an earthfill dam

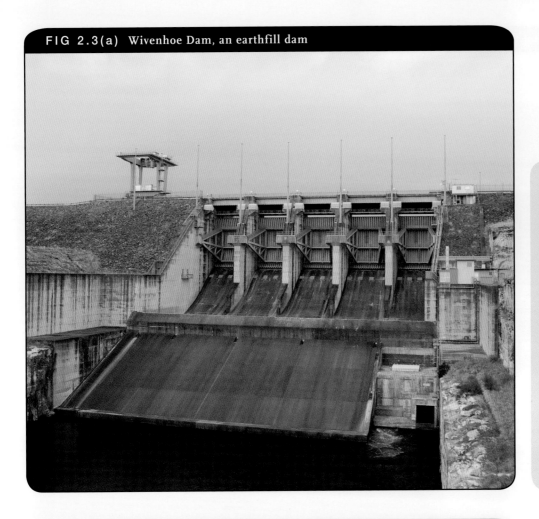

Wivenhoe Dam is built on the Brisbane River and ensures a safe and reliable water supply to south-east Queensland. It consists of an earth and rock embankment 2.3 kilometres long and 50 metres high and has a capacity of 1 150 000 megalitres. It also pumps water into the Splityard Creek Dam for the production of hydro-electricity at times of peak demand.

FIG 2.3(b)   Dartmouth Dam, an earthfill dam

Dartmouth Dam stores water for irrigation, domestic and stock use in Victoria and New South Wales. It is also used to generate 150 MW of hydro-electricity. In dry seasons it supplements releases from Lake Hume and increases supplies to the Murray River system. It is the highest dam in Australia at 180 metres and has a capacity of 4 000 000 megalitres. The embankment is made up of 14.5 million cubic metres of earth and rock fill.

FIG 2.4(a) Warragamba Dam, a concrete gravity dam

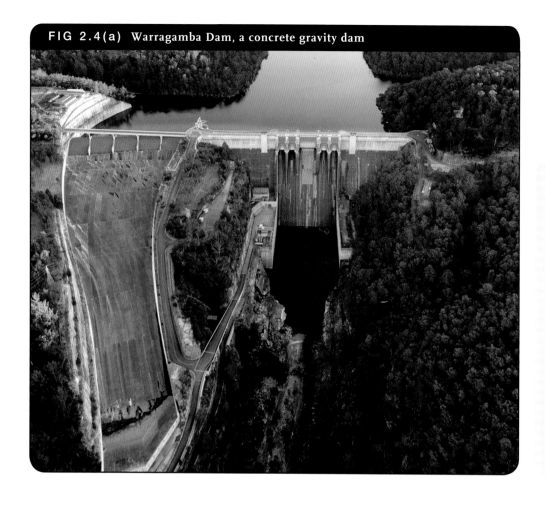

Warragamba Dam is the main storage for Sydney and its surrounds, and is one of the largest domestic water supply dams in the world. It is the highest concrete gravity dam in Australia at a height of 142 metres. It covers an area of 7500 hectares, has a capacity of 2 057 000 megalitres, and has a primarily protected catchment area of 9000 square kilometres.

FIG 2.4(b) Victoria Dam, a concrete gravity dam

Victoria Dam was built to supply water to the adjacent hills suburbs of Lesmurdie and Kalamunda. It also supplements Perth's water supply during times of high demand. The dam has a catchment area of 37 square kilometres and a capacity of 9500 megalitres. It contains 134 500 cubic metres of concrete with a wall height of 52 metres and length of 285 metres.

Catchment areas

The size of a catchment area varies considerably with the amount of rainfall available in the area and the demand for water by the population which the reservoir serves. In Sydney, for example, the average consumption per head. per day is 515 litres. This figure is based on industrial and commercial use as well as domestic requirements However, this figure is calculated over the whole year and in the peak demand months of December, January and February the daily consumption may rise to 1000 litres per head per day, although recommended water saving strategies and water use restrictions may greatly reduce this figure.

To store this amount of water requires an extremely large storage dam or a series of dams and reservoirs. Correspondingly large catchment areas are required to maintain the levels in the storage dams. The control of these vast areas is the responsibility of the water control authorities who employ large numbers of rangers to patrol and guard against possible pollution from any source, whether human or animal. They also inspect the area for erosion and bushfires.

The capacities stated above refer to dams when they are full. Due to inconsistant rainfall and changing weather patterns, dams are rarely full and, in times of severe drought, can fall to alarming levels. To ensure a constant supply, alternative methods of water supply are used to supplement stored water, along with the imposition of various levels of restrictions on water use in times of drought. In times of flood, the dams spill over into downstream rivers and eventually out to sea.

Agriculture is restricted in prescribed catchment areas. Any remaining privately owned properties are eventually purchased by the authority to ensure that pollution caused by agricultural chemicals, animal faeces and so on, is reduced. These areas eventually become havens for native animals and the patrolling rangers are watchful for anything that could become a source of pollution.

Service reservoirs

Reservoirs are provided by the water authority to store treated or purified water before it is distributed to the consumer. They are generally installed within the area served, as close as possible to the point of final consumption, and are usually constructed of either concrete or steel (Figure 2.5).

The function of a service reservoir varies considerably from one water supply authority to another. It may be used for any or all of the following purposes:

- to retain a reserve of potable water for use if mechanical, electrical or mains failure prevents the operation of pumping equipment
- to supply a reserve of water in areas where a substantial variation in demand occurs
- to provide a constant head on systems
- to act as a break tank (which reduces the operating head) in low-lying areas supplied directly from impounding reservoirs.

FIG 2.5 Steel and concrete service reservoirs

Service reservoirs are located as near to the centre of the distribution area as possible and at sufficient elevation to provide satisfactory water pressure over the entire area supplied. They are usually enclosed (roofed) to prevent contamination of the water by either atmospheric or animal pollution. Pressure created by the height can be calculated by using the formula

$$P = H \times 9.81.$$

This was previously discussed in Chapter 1.

In areas where it is difficult to obtain sufficient height or head, service reservoirs are erected on towers. These water towers are generally smaller in capacity than those constructed at ground or below ground level (Figure 2.6).

FIG 2.6 Modern, all-steel water tower

IMPURITIES IN WATER

All water, irrespective of its source, contains some impurities. These impurities enter the water as it falls through the atmosphere as rain, or as it flows over the ground. Underground or subsurface water also contains impurities due to the dissolving of minerals from the earth as it passes through.

Water may also be contaminated or polluted after it has been collected, treated and distributed. This may be caused by a break in the distribution pipework or, more commonly, by the faulty installation of appliances or equipment.

Generally the operative plumber is not concerned with the quality of the drinking water for which he or she is installing pipework or other equipment. This is the responsibility of the various state departments of health, who publish minimum standards of safety for water distributed within their areas of jurisdiction. However, it *is* up to the plumber to ensure that the integrity of the water is not compromised, particularly by contamination through backflow.

The numerous water supply authorities throughout Australia collect and test samples of water to ascertain the type and concentration of impurities that may be present and to determine the amount of treatment necessary to maintain the standards of drinking water required by the health authorities.

In areas where distributed water requires additional treatment, other than that carried out by the water authority, such as water softening, the plumber may be requested to install small package treatment plants to improve the water quality. This may also occur in areas where water is drawn from sources other than municipal dams, as it often contains minerals that makes the water hard. (For further information regarding the installation of package treatment plants refer to Domestic Water Softeners in the section on Water Treatment.)

Types of impurity

Impurities in water can be divided into two main groups: inorganic and organic. Inorganic impurities are of a chemical nature and are dissolved in the water. These include calcium or magnesium salts which cause hardness. Inorganic impurities may also be insoluble and remain in the water in suspension, such as clay and sand particles. The insoluble type may settle after a period of storage.

Organic impurities (bacteria) are not all detrimental to humans; in fact, some may even be beneficial. They can be either alive or dead and include such things as plankton (a minute form of animal life) and algae. The presence of either of these impurities in water can sometimes be indicated by unusual odours or discolouration.

Odours in water are caused in the majority of cases by either or all of the following:

- dissolved gases
- decomposing organic matter
- industrial wastes and discharges
- chlorine, either residual or combined with other substances.

Reliance on odours or discolouration is not a safe method of detection however, as both depend on the amount or concentration present. Pathogenic (disease producing) bacteria may be introduced into water when it passes over land that has been fertilised for agricultural purposes or has been fouled by human or animal faeces.

If water intended for human consumption is suspected of containing impurities of any kind, samples should be submitted to either the water supply authority or health authority for thorough chemical, physical or bacteriological examination.

Turbidity, or a muddy, hazy appearance, is produced when light is refracted by particles of sand or silt that are held in suspension in the water. *Colour* in water is often caused by suspended matter, such as clay or silt, or tannin from decaying leaves or other vegetable matter. Discolouration of water by dissolved minerals such as iron oxide occurs in some areas, although this is not a common problem because of the care taken in selecting catchment areas.

Hardness occurs in water that has passed through geological strata containing a high concentration of mineral salts. It is extremely difficult to produce a soap lather in hard water. Hard waters are classified into two groups: temporary hard water and permanent hard water.

Carbonate hardness is caused by the presence of carbonates and bicarbonates of calcium and magnesium. This type of hardness is often referred to as temporary because it may be removed simply by boiling the water. When the water is boiled, carbon dioxide gas is driven off, leaving the hardness-producing salts to precipitate out in the form of crystals that are removed by filtration. This process is only practical for small volumes of water. In treatment plants handling large volumes, the ion exchange or zeolite process is used. The process is effective for the removal of both temporary and permanent hardness, and operates by exchanging the calcium and magnesium for sodium.

Permanent or non-carbonate hardness is caused by the presence of sulphates and chlorides of calcium and magnesium which cannot be removed by boiling the water. The addition of chemicals such as sodium carbonate, lime or caustic soda is required to neutralise this type of hardness.

In general, the mineral content of water does not create problems in water supply, providing the concentration of these minerals remains small. Mineral concentrations are usually expressed in milligrams per litre (mg/L) or in parts per million (ppm). Because of the solvent characteristics of soft water, it may be possible to find traces of copper or zinc in potable water. This is removed from the inside of pipework or storage vessels as the water passes through. The amount contained depends on the acidity or alkalinity of the water.

The acidity or alkalinity of a substance is usually referred to as its pH (power of hydrogen). The standard pH scale ranges from 0 to 14 and the number given to a substance within that range indicates its acidity or alkalinity. Neutral substances—that is, neither acid nor alkali—have a pH of 7, those below 7 being acid, and above 7 alkaline. For water treatment purposes, those waters with a pH reading falling between 6 and 8 are regarded as being neutral.

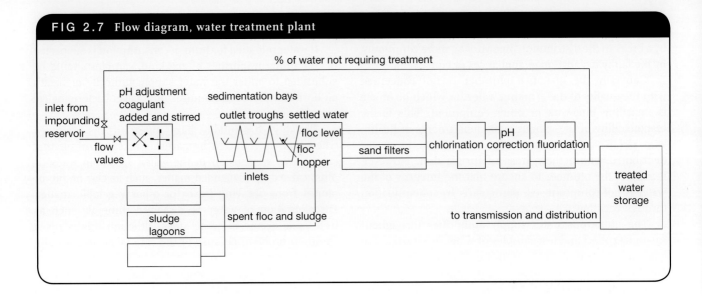

FIG 2.7 Flow diagram, water treatment plant

WATER TREATMENT

Water treatment, carried out by a water authority in a modern treatment plant, is a complex operation requiring accurate control and constant testing and chemical analysis. This is to ensure that only water complying with the minimum standards laid down by the relevant state health departments is delivered to the consumer. In the operation of a water treatment plant it is necessary to understand the following processes (Figure 2.7).

1   Coagulation is the chemical reaction between a coagulant—generally aluminium sulphate (alum) or ferrous sulphate—and the water to be treated. For a satisfactory reaction the water must be slightly alkaline. In situations where the raw water has insufficient alkalinity, the pH is adjusted by introducing slaked lime or soda ash into the water.

2   Flocculation is the process brought about by stirring the coagulant into the raw water. It results in minute gelatine-like particles (floc) forming around the turbidity-producing substances. The floc particles grow bigger and settle, carrying any suspended solids and bacteria to the bottom.

3   Sedimentation (sometimes referred to as 'settlement') takes place after flocculation, the water being allowed to stand in large tanks or bays for approximately two to six hours. This permits the floc and colloidal matter (finely divided solids) to settle out.

4   Filtration to remove the finely divided solids and floc consists of passing the water through beds of specially selected and graded sands. There are two basic methods of sand filtration: slow and rapid. Slow sand filtration is a process requiring an extremely large area of sand bed through which the water is allowed to percolate. Water which has not been pretreated may be passed through slow filters, although this method is unable to handle high turbidity as it tends to reduce the filtering capacity of the bed more quickly.

Rapid sand filtration requires careful pretreatment of the water to be effective as a filtering method; coagulation and flocculation are of paramount importance as the floc film that settles on the bed plays an important part in the filtering process.

Filtration through sand is also an effective method of removing bacteria.

5   Sterilisation is required to remove any harmful bacteria that may be present. Chlorine, either in solution or as a gas, is introduced into the water with accurate measuring equipment. Chlorine is an efficient disinfectant and sterilisation is considered effective if, after a contact period of thirty minutes, there is a satisfactory chlorine residue in the water. The amount of residual chlorine in the water is determined by either a standard chlorine colour test or by laboratory analysis. UV sterilisation is also employed in many treatment plants usually in conjunction with chlorination. Ozone may also be injected via a pressure system to assist in the sterilisation process.

6   pH correction is the process of adjusting the acid/alkali balance of the water. After the ferrous sulphate that is required for coagulation is added, the water tends to be acidic. This excess acidity is then neutralised by the addition of small amounts of sodium carbonate, which raises the pH to approximately 7.5, which is slightly alkaline. This reduces the ability of the water to attack and corrode the interior of pipes and storage vessels.

7   The fluoridation process consists of adding minute doses of fluoride compound to the water as it passes through the plant. The ultimate concentration is approximately one part per million and it is claimed that at this concentration it reduces dental cavities in children by 65 per cent. The chemicals used are extremely dangerous and complex measuring equipment is used to ensure that the correct amount of sodium silicofluoride is introduced into the water supply.

Further information on water treatment in your area can be obtained from your local water authority.

Home treatment

Where the water supplied for drinking purposes has not been fully treated by the water authority for economic reasons, or has received no treatment—as in rural water supplies drawn from local resources—it is necessary for the consumer to treat the water before use. Some industrial and commercial processes also require water with specific qualities and this is achieved by supplementary treatment before use.

Domestic water treatment may consist of simple operations such as filtration or a more complex treatment such as softening.

Filters

There are numerous filters available for domestic use. They can remove impurities such as sand, silt, rust, scale and algae, and numerous others that can be present in potable water. Filters can also remove unpleasant tastes and odours from water. These filters are small, compact and easily installed, either at a single outlet or in a position where the entire water supply to a residence requires filtering (Figures 2.8 (a,b and c)).

Cartridges for filters are replaceable. The type most commonly used to remove sediments or colloids from water is usually constructed from pleated polyester fabric or resin-impregnated paper; those used for the removal of tastes and odours contain activated charcoal. Filters are also available that can effectively remove combinations of impurities, or filters may be installed in tandem when combination types are unavailable or unsatisfactory.

Where filters are installed, the flow of water in the pipeline is restricted and a corresponding pressure drop occurs. For this reason, adequate compensation should be allowed when pipe sizing an installation for this apparatus. For example, a filter installed in a water service delivering water at a pressure of 276 kPa would reduce the outlet

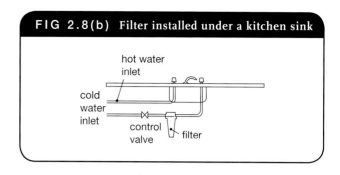

FIG 2.8(b)   Filter installed under a kitchen sink

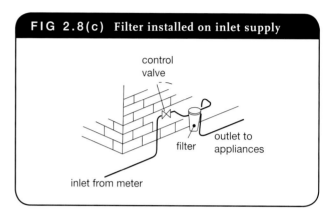

FIG 2.8(c)   Filter installed on inlet supply

pressure to approximately 255 kPa, while maintaining a discharge through the filter of some 45 L/min.

Most filters don't remove fluoride. Where fluoride removal is required, reverse osmosis is usually employed. This is similar in style to a cartridge set but with the addition of a final filter and UV sterilisation (Figure 2.9(a)).

Desalination

In an environment where weather patterns are changing and rainfall is not as reliable as it once was, alternative methods of water supply are becoming increasingly important. One of these alternate supplies is desalination. Desalination consists of taking water from the sea and running it through a process to purify it and remove the salt. The process requires varying amounts of energy and production of waste water (high in salt concentration) dependent upon the method used for desalination. The running of the plant is often energy intensive and costly compared to other methods, but with advances in technology and renewable energy, it may become a viable alternative in future-proofing our water supplies.

Reverse osmosis (RO) is one of the more popular processes used for desalination. It essentially consists of passing pressurised salt water through a membrane (Figure 2.9(b)). As it passes through the membrane it separates the water from the salt, leaving behind a high saline solution. Only a small portion of the water drawn in passes through the membrane and the highly saline solution left behind is discharged to waste. Before passing through the RO membrane it is firstly pretreated through filtration to remove solids and suspended matter. After the reverse osmosis process it is then further treated to balance the pH and chlorinated. Fluoride may also be added at this stage.

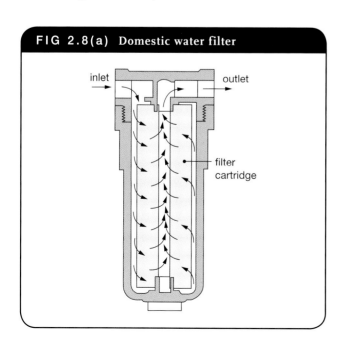

FIG 2.8(a)   Domestic water filter

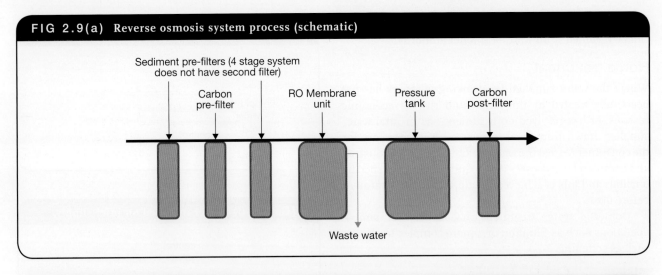

FIG 2.9(a) Reverse osmosis system process (schematic)

Sediment pre-filters (4 stage system does not have second filter)

Carbon pre-filter

RO Membrane unit

Pressure tank

Carbon post-filter

Waste water

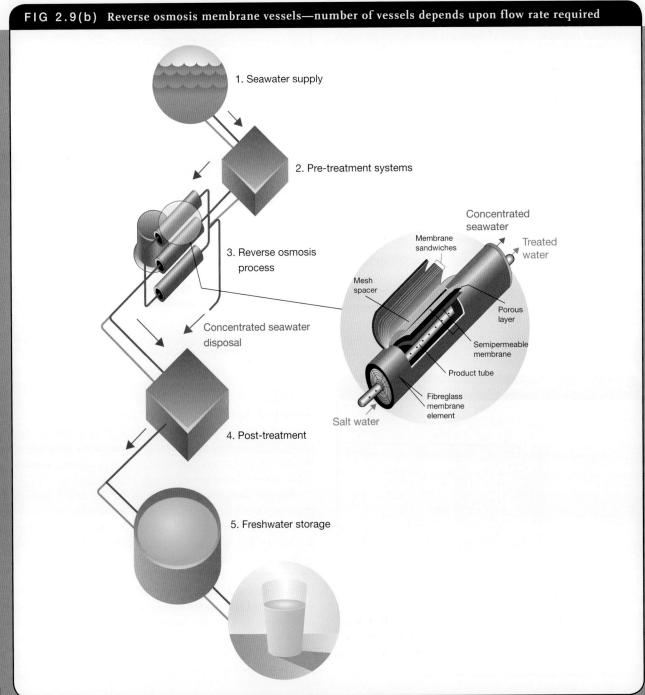

FIG 2.9(b) Reverse osmosis membrane vessels—number of vessels depends upon flow rate required

1. Seawater supply

2. Pre-treatment systems

3. Reverse osmosis process

Concentrated seawater disposal

4. Post-treatment

5. Freshwater storage

Concentrated seawater

Treated water

Membrane sandwiches

Mesh spacer

Porous layer

Semipermeable membrane

Product tube

Fibreglass membrane element

Salt water

SUSTAINABLE PLUMBING: DESALINATION

A major concern about desalination is the high energy input and resultant greenhouse gas emissions associated with it. This is often offset by providing greenhouse-friendly methods of energy production, such as wind generators or solar energy, but it still remains an issue for many. Another concern is the effect the high salt waste discharge can have on the environment, in particular the local marine environment. To combat this, water authorities constantly monitor the waste discharge to ensure that it does not have a detrimental effect on the environment.

Most states and territories in Australia either operate or are building desalination plants to supplement their water supply. Sydney can provide up to 15 per cent of its water supply through desalination and Perth can provide up to 17 per cent—and is aiming to provide almost half of its drinking water by this method. Adelaide's desalination plant is designed to provide 50 per cent of its drinking water. For cities such as Perth and Adelaide, where dam water reserves are not as significant as for other states, sources such as desalination are important to ensure a continuous supply of clean drinking water.

UNDERGROUND WATER

There are vast reserves of water below ground stored in pervious layers of sand, sandstone or limestone that are caught between various layers of non-pervious structures such as rock or clay. Water is usually drawn from the unconfined aquifer closest to the ground surface with its upper surface the water table. There are also considerable deeper reserves in confined aquifers where the water is under pressure. These are sometimes referred to as artesian waters.

Groundwater usually contains higher concentrations of natural dissolved materials than surface water. It usually reflects the composition and solubility of the earth that the groundwater is in contact with and the length of time that it has been in the subsurface. These minerals are mostly harmless although the water is often 'hard' due to the mineral content and may taste different to dam-supplied water, again due to the mineral content. Generally, underground water is considered clean as infiltration through the soil is a form of water treatment; but it should always be tested, as there is always the possibility of contamination from human activity (Figure 2.10(a)).

Groundwater is often the main source of supply for rural properties for domestic, stock and irrigation purposes and is often used without further treatment as its quality is of a relatively high standard. Activity in the surrounding area needs to be monitored so as not to jeopardise this important resource and maintain its quality. Many coastal properties also use ground water for irrigation purposes, even though they are connected to a municipal supply.

A number of activities that can pose threats to underground water quality may include but are not limited to:

* landfill waste disposal infiltration
* septic waste infiltration
* industrial activities
* mining activities
* treatment and distribution.

Depending on the season, between 35 and 50 per cent of Perth's public water supply comes from treated groundwater. The groundwater is pumped from aquifers to treatment plants (Figure 2.10(b)), where it is treated to make it suitable for drinking before being distributed to customers. The common water treatment processes of aeration, settling of suspended solids, filtration and chlorination as well as resin magnetic ion exchange are used.

RECYCLED WATER

Recycled water is generally sourced from sewage wastewater or captured stormwater. This water would otherwise be discharged as waste, often to rivers or the sea. Stormwater is reliant on weather patterns and rainfall, whereas sewage is a constant supply, with a high percentage of the drinking water that enters a property returning via the sewerage system. The flowchart in Figure 2.11(a) outlines the recycled treatment process. Recycled water is treated to be 'fit for purpose'. Intended uses include commercial irrigation and industrial processes as well as domestic uses, such as garden or lawn irrigation, toilet flushing and car-washing. With further treatment it can also be made fit for drinking, although this still

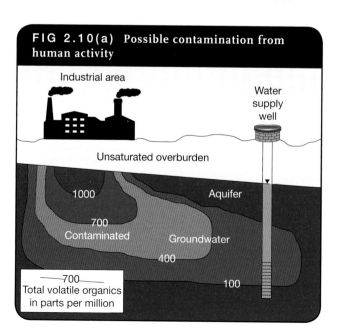

FIG 2.10(a) Possible contamination from human activity

Industrial area

Water supply well

Unsaturated overburden

1000

700
Contaminated

Aquifer

Groundwater

400

100

———700———
Total volatile organics in parts per million

FIG 2.10(b) Flowchart of groundwater treatment process at Wanneroo Ground Water Treatment Plant (GWTP), Western Australia

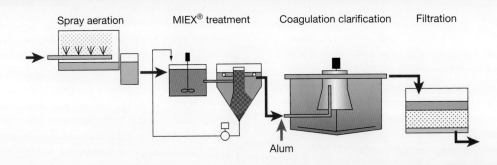

Spray aeration   MIEX® treatment   Coagulation clarification   Filtration

Alum

FIG 2.11(a) Recycled treatment process

Sydney
WAT≥R

Recycled water treatment processes

Untreated wastewater
Wastewater going to treatment plants is more than 99% water. It comes from bathrooms, kitchens, laundries and businesses.

Primary treatment
This step removes large solid particles from wastewater using screeening, grit removal and sedimentation.

Secondary treatment
Smaller particles and dissolved pollutants remain after primary treatment. 'Good' bacteria feed on and clean up organic particles and nutrients. Other typical treatment processes include aeration, settling, clarification and chemical treatment.

Tertiary treatment
Deep sand beds are used to filter out nearly all remaining organic particles and suspended material.

Advanced treatment
This step may include microfiltration, ultrafiltration and reverse osmosis. These filters are so fine they can trap particles smaller than a millionth of a metre.

Disinfection
This is the last step for all types of recycled water.
The water is disinfected by chlorination, ultra violet light, or both processes.

Recycled water—ready for its intended use.

This type of recycled water can be used for controlled irrigation.

This type of highly treated recycled water is suitable for many uses, including irrigation, industrial processes and for watering gardens and flushing toilets.

This type of highly treated recycled water is for specialised uses such as some manufacturing processes and for river flows.

Guidelines for using recycled water in residential areas

Recycled water can be safely used to:

- wash the car
- flush the toilet
- fill ornamental ponds
- water most garden plants
- fight fires
- wash laundry in a washing machine.

Do **not** use recycled water for:

- drinking
- personal washing such as baths, showers, bidets and hand basins
- recreation involving water contact
- filling swimming pools and spas
- cooking and other kitchen purposes
- household cleaning
- watering fruit and vegetables that are to be eaten raw or unpeeled.

carries a stigma for some. The treatment processes are a combination of those used for wastewater and drinking water treatment, except that there are obviously more contaminants to remove from wastewater than dam water.

Chlorination is heavier for recycled water due to its source and is often coupled with UV sterilisation. Figure 2.11(b) shows Sydney Water's Rouse Hill treatment plant layout and the scale of its processes.

FIG 2.11(b) Rouse Hill Sewage Treatment and Recycled Water Plant (Sydney Water)

FIG 2.11(c) Recycled and drinking water meters

FIG 2.11(d) Recycled water hose tap

Recycled water components are identified by several means. Meters are colour-coded (usually purple or lilac) and have different sized connections to drinking water meters (Figure 2.11(c)). Hose taps are identified by colour, have a removable handle, and a left-hand thread on the outlet, and are signposted as not suitable for drinking (Figure 2.11(d)). Tanks storing water other than municipal drinking water are also signposted.

Re-use of greywater

Greywater is water from waste fixtures in bathrooms and laundries (Figure 2.12). It does not include waste from kitchens or blackwater from soil fixtures. It can be re-used either treated or untreated within certain limitations. Devices can be as simple as a diversion unit to divert washing water or as sophisticated as a commercial treatment plant.

Diversion of greywater can be used for sub-surface irrigation. As there is the likelihood that it may contain pathogens, care should be exercised when using greywater. Do not allow it to pond and do not store greywater for longer than 24 hours; it is also not suitable for sprinklers as the fine spray can drift.

Treated greywater may be used for flushing toilets and laundries—basically the same uses as recycled water, dependent upon the level of treatment. Treatment may include aeration, filtration, ultrafiltration, membrane bioreactor, UV disinfection and chlorination. Check with manufacturers as to the specifications of their system and its suitability for your intended purpose. Greywater is also heavily regulated and you need to check with your local council and health and water authorities as to their requirements and application procedures.

Sewer mining

Sewer mining consists of tapping directly into the sewer and diverting it to a treatment plant where the sewage is made fit for purpose and then returning the by-product back to the sewer.

Treatment can consist of filtration, microfiltration, membrane bioreactors, reverse osmosis, UV disinfection and chlorination. As with recycled water, the permitted uses depend on the level of treatment (made up of a combination of the processes mentioned above), but are typically irrigation and toilet flushing. All components need to be identified as recycled water. One advantage of sewer mining is that it is at the source of intended re-use thereby negating the need for transporting the treated water over distances.

It is suitable for both small applications such as individual buildings, and large applications such as large-scale irrigation or new developments. Due to the localness of the process it is usually privately run. Figure 2.13(a) outlines the process of one such large application used for irrigation. Mined sewer water is particularly suitable for irrigation (Figure 2.13(b)) as it is high in nutrients, thereby minimising the amount of fertiliser required by up to 90 per cent.

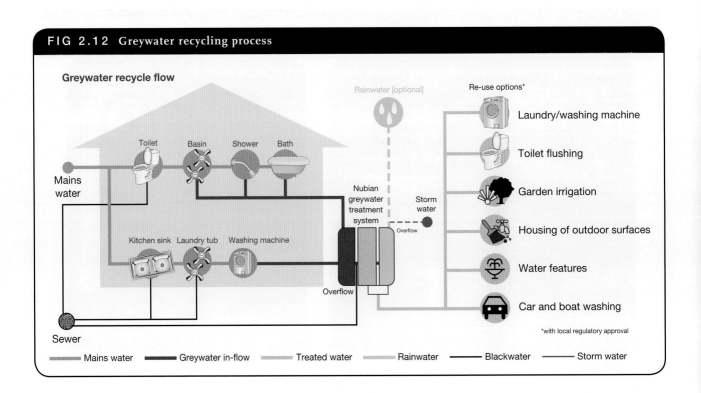

FIG 2.12 Greywater recycling process

Greywater recycle flow

Rainwater [optional]

Re-use options*

Toilet   Basin   Shower   Bath

Mains water

Nubian greywater treatment system

Storm water

Overflow

Kitchen sink   Laundry tub   Washing machine

Overflow

Sewer

Laundry/washing machine

Toilet flushing

Garden irrigation

Housing of outdoor surfaces

Water features

Car and boat washing

*with local regulatory approval

Mains water ——— Greywater in-flow ——— Treated water ——— Rainwater ——— Blackwater ——— Storm water

FIG 2.13(a)   Typical sewer mining process

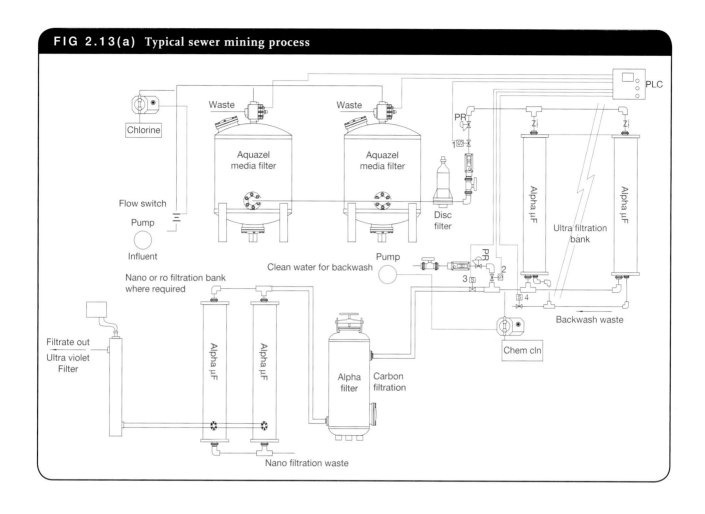

FIG 2.13(b)   Samples of water before, during and after treatment from the Pennant Hills Golf Course sewer mining treatment plant

Stormwater and rainwater re-use

The rainwater that lands on your property is an obvious alternative water supply. However, consideration needs to be given to the surfaces that it is collected on and the surrounding environment.

Rainwater collected from your roof usually needs minimal treatment to keep contaminants out of the tank. Rainwater can be used for most applications but should not be used for drinking and food preparation where municipal water is available, as the quality of rainwater cannot be guaranteed. It is recommended that tanks should be cleaned periodically (bi-annually) and also disinfected if used for drinking water. Roofs and gutters should be cleaned every six months. Figure 2.14(a) demonstrates a total rainwater harvesting system.

The surface that stormwater is collected over is not as clean as roof water because it includes run-off from the surrounding area and can include all sorts of contaminants (e.g. dog faeces). With this in mind, the re-use of stormwater should only be used for low-risk activities such as irrigation and flushing toilets. Of course if you install a treatment system, as in commercial installations, this increases the usability of the collected water. Figures 2.14(b) and (c) illustrate a commercial stormwater harvesting and re-use system utilising filtration and sedimentation. Primary treatment where needed is achieved through a gross pollutant trap, which can remove particles up to 150 um in size, and secondary treatment through a filter, prior to entry into the re-use system.

DOMESTIC WATER SOFTENERS

Hard water is water that contains dissolved calcium and magnesium. If the calcium and magnesium are removed, the water becomes soft.

Ion exchange involves the removal of 'hard' ions from the water and substituting other 'soft' ions in their place. When most materials dissolve in water they 'ionise', or form ions, single atoms, or in some cases a small group of atoms, that carry a very small electrical charge that can be either positive or negative. In water, there is no net charge because the positive charge is always equal to the negative charge.

Water softening by ion exchange is not a new process. The water softening properties of natural materials, certain clays and sands, have been studied for more than 100 years. These natural exchanger substances are called zeolites. Studies of zeolites and their exchange properties led to the development of softening units.

FIG 2.14(a) Rainwater harvesting system (domestic)

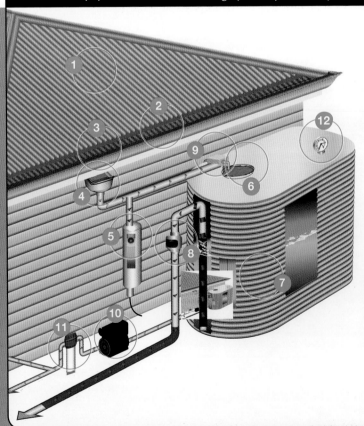

1. Check Roof Surface is suitable for collecting quality rainwater.

2. Install Gutter Mesh to prevent leaves and debris from blocking gutters.

3. Fit Gutter Outlets from the underside of the gutter to prevent obstruction of water flow.

4. Fit Rain Heads to downpipes to prevent blocking. Rain heads deflect leaves and debris and keep mosquitoes out of pipes that hold water ('wet' systems).

5. Install First Flush Diverter/s to help prevent the most contaminated rainwater from entering the tank.

6. Ensure a Tank Screen is installed at tank entry point to keep mosquitoes and pests out.

7. Choose a Water Tank. Consider annual rainfall, roof catchment area and water usage when determining its size.

8. Attach Insect Proof Screens or Flap Valves to the end of all pipes to the tank screen to keep mosquitoes and pests out and ensure the tank is vented properly. Install an Air Gap to Tank Overflow Outlets to prevent stormwater backflow into your tank.

9. Utilise a Tank Top Up system (if required) to automatically top up the tank with mains water when levels fall to a designated minimum level.

10. Select a Pump System (if required) to distribute water for use inside or outside the home.

11. Rainwater Filter. Fit a purpose designed rainwater filter after the pump to help reduce residual sediment, colour and odour.

12. Water Level Indicator. Install a tank gauge water level indicator to your rainwater tank to monitor your water level and usage.

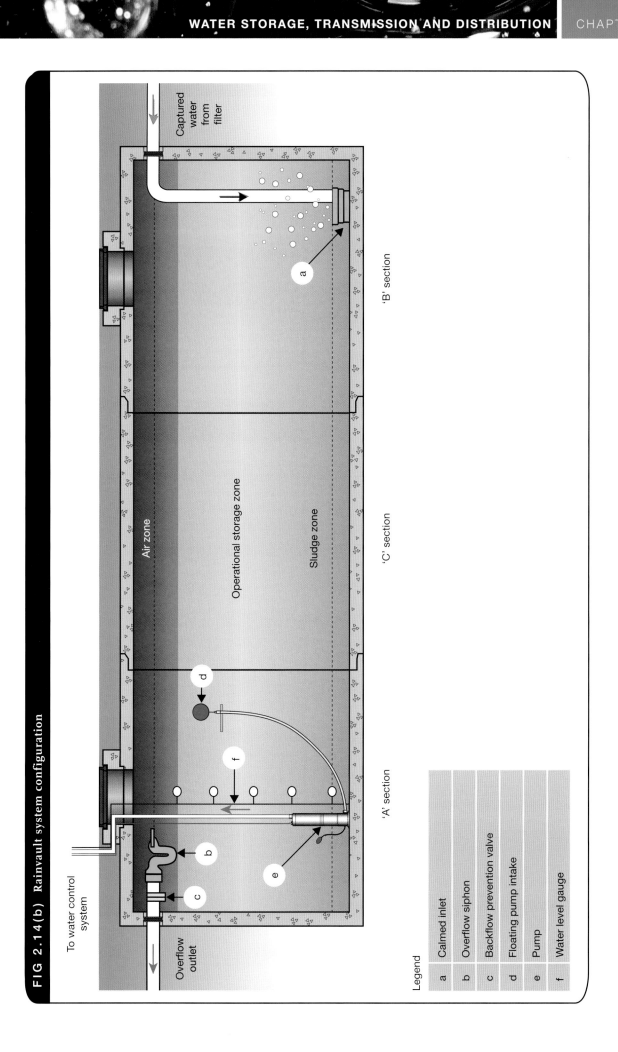

FIG 2.14 (b) Rainvault system configuration

Captured water from filter

'B' section

Air zone

Operational storage zone

Sludge zone

'C' section

'A' section

To water control system

Overflow outlet

Legend

a	Calmed inlet
b	Overflow siphon
c	Backflow prevention valve
d	Floating pump intake
e	Pump
f	Water level gauge

FIG 2.14(c) Rainvault system being installed

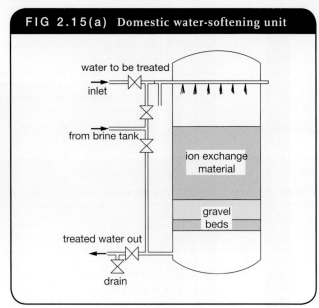

FIG 2.15(a) Domestic water-softening unit

Synthetic ion exchange agents are known collectively as 'ion exchange resins'. These are porous, bead-like materials consisting of many tunnels and cavities that increase their surface area and therefore increase the ion exchange potential of the substance.

In water softening by ion exchange, 'hard' ions (calcium and magnesium) are extracted from the water and 'soft' ions (sodium) are substituted. To enable this, the ion exchange resins are 'preloaded' with sodium chloride (common salt). The water is softened as it passes through the layer of resin. In this way the calcium and magnesium ions are removed, rendering the water soft as it leaves the resin layer.

As in every other natural process, ion exchange cannot continue indefinitely. As the resin becomes saturated with calcium and magnesium, it is unable to remove any more and replace it with sodium because all the sodium ions have been exchanged.

The softening capacity of the resin can be restored by a process known as 'regeneration'. In this process, the flow of water through the softener is reversed and a salt solution is fed through the resin bed for twenty or thirty minutes. The sodium ions in the salt solution scour the calcium and magnesium ions from the resin. As a result the resin is again charged with sodium ions and can be used again for water softening (Figure 2.15(a)).

Water softening plants are available that carry out the regeneration process either manually or automatically. The size of these package treatment plants varies considerably, from small compact units designed to treat sufficient water for a normal domestic cottage (Figure 2.15(b)) to units capable of use in factories for specific treatment processes.

FIG 2.15(b) Method of installing domestic water-softening plant

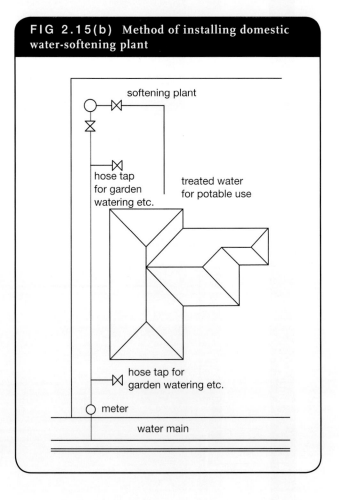

WATER TRANSMISSION AND DISTRIBUTION

Transmission of water, like its collection and treatment, is usually the responsibility of the water supply authority and only on rare occasions does the plumber become involved in this area of operation. Transmission pipework conveys the collected and treated water to the distribution system. The location of the water collection facility and its elevation above the area where the water is to be distributed determine the length and type of transmission system used (Figure 2.16). If, for example, the catchment area storage dam and treatment works are at a higher elevation than the distribution system, the water may be transported through open channels by gravity. In situations where the reverse occurs or to prevent contamination, it is necessary to supply water to service reservoirs by pumping it through large diameter pipelines (Figure 2.17).

Open channels or canals are generally constructed by cut and fill methods—openings are cut through hills and the waste that is removed is used to fill valleys so that the channel follows the hydraulic gradient (Figure 2.18). Pipelines used for this purpose generally follow the topography of the ground over which they pass and in some instances rise above the hydraulic gradient. The size of these transmission conduits is determined by the capacity of the distribution system, the current and future demands of the area to be served, and other economic considerations.

Open channels have distinct advantages and disadvantages when compared with the conventional pipe. The principal disadvantages are the amount of water lost by evaporation and the added risk of contamination, especially in populated areas; as such, open channels are mainly used for supplying water to hydro-electricity generators rather than for drinking water. The main advantage with open channels is that their construction costs are much lower compared to pipelines, provided that cut and fill methods are used during construction.

Distribution systems are the part of the water supply system that extends from the service reservoir to the streets of a community. From there, the individual services to properties are taken. The system is made up of feeder mains, secondary mains and reticulating services.

It is here that the operative plumber is sometimes involved. Although the mains laid in our streets are generally installed and maintained by the water authority, authorised plumbing contractors can be engaged by property developers to install water mains in new subdivisions or by the water supply authority itself on a subcontract basis. This often occurs in areas where the water supply is provided by a small authority that does not have sufficient or qualified staff to carry out this type of work.

There are two distinct patterns used in designing a distribution system. These are the tree pattern (Figure 2.19) and the grid pattern (Figure 2.20). There are also variations on these two basic patterns, an example being the grid with the feeder main looped around a densely populated area or an industrial area (Figure 2.21). The tree configuration of distribution pipework is used where the development of an area is in one direction only (Figure 2.22). In this situation the feeder is usually sized to allow for future development or later inclusion in a grid system.

FIG 2.16 Brick aqueduct in use until 1995, now heritage listed, Sydney

FIG 2.17 Large diameter steel pipelines used to convey potable water

FIG 2.18 Cut and fill construction of open channel

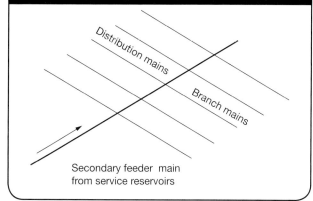

FIG 2.19 Tree pattern layout of water mains

Distribution mains

Branch mains

Secondary feeder main
from service reservoirs

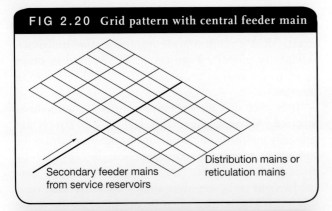

FIG 2.20 Grid pattern with central feeder main

Secondary feeder mains from service reservoirs

Distribution mains or reticulation mains

FIG 2.21 Grid pattern with feeder main looped around area of high population density or industrial activity

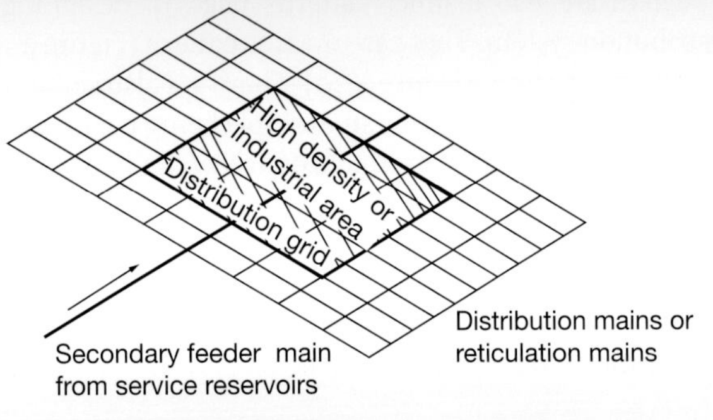

High density or industrial area

Distribution grid

Secondary feeder main from service reservoirs

Distribution mains or reticulation mains

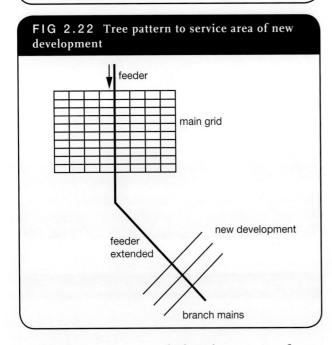

FIG 2.22 Tree pattern to service area of new development

feeder

main grid

feeder extended

new development

branch mains

Distribution mains are laid in the majority of cases using the grid system. In this system the water enters any particular main from two directions. This ensures that if a breakdown occurs, the continuation of supply is maintained from at least one direction.

DISTRIBUTION MATERIALS

In the past, cast iron was used exclusively for water mains and other apparatus in the distribution system. Modern manufacturing techniques improved the quality of cast iron pipes and fittings and this much improved material is still in service in existing older water mains but has been replaced by ductile iron for new construction and repair work.

There are several other materials used for water mains in Australia, each with specific characteristics and laying techniques. The type of materials used for water mains is dictated by installation cost, geographic location, atmospheric conditions, topography of surrounding countryside, aggressiveness of soils and many other factors.

Materials other than ductile iron used for water distribution in Australia are: steel, glass-reinforced plastic (GRP) polyvinyl chloride (PVC) and polyethylene.

Cast iron pipe

The early cast iron pipe was cast vertically in sand moulds and produced an extremely heavy wall pipe, which was porous in structure and of doubtful tensile strength. This manufacturing technique was superseded by centrifugally casting the iron in steel moulds to produce what was referred to as spun cast iron. Spun cast iron possessed a fine grain structure and increased tensile strength. Because of the reduced grain size, the wall thickness of the pipe was substantially less than the earlier vertically cast type. This dramatically reduced the overall mass of the material, thereby reducing transport and handling costs and making it easier to install. Grey cast iron is no longer manufactured for use as water mains and has been replaced by ductile iron.

Ductile iron

Ductile iron is manufactured in a similar manner to spun cast iron pipe except that magnesium is added to the molten metal before casting. The addition of the magnesium removes the sulphur from the casting metal and produces a different grain structure to grey cast iron. This alteration of grain structure enables the pipe wall thicknesses to be reduced and also increases the ductility of the pipe, hence its name.

Ductile iron pipe is suitable for installation in areas where the pipe may be subjected to extreme stress such as crossing bridges, under high traffic areas or in positions where the pipe may be subjected to external damage. The light weight and improved ductility dramatically reduce handling and transportation costs and also reduce the risk of damage to pipe coatings and linings.

Ductile iron coatings and linings

All ductile iron pipes and fittings are protected against external corrosion by various protective paint, polyurethane or epoxy coatings selected to suit the soil conditions. Care should be exercised when handling pipes coated with these materials to ensure that the external coating is not damaged. In areas where the surrounding soil is 'aggressive', additional protective coatings may be required to extend the underground life of the pipes and fittings. Additional protection can be provided by sleeving the installation with polyethylene.

Internal corrosion protection of water mains is effected by the application of a cement mortar lining to both pipes and fittings. This lining is applied in accordance with the relevant national standard (e.g. AS/NZS 2280). The thickness of this lining varies with the diameter of the pipe. Cement-lined pipes should be handled with extreme care to minimise the chances of damaging the pipe lining.

Ductile iron joints

Rubber ring joints consist of a specially designed socket that accommodates an elastomeric (EPDM) gasket (Figure 2.23(a)). The gasket should be wiped clean, flexed as shown in Figure 2.23(b) and then placed in the socket with the rounded bulb of the gasket to the back of the collar. The groove in the gasket must be located on the retaining bead in the socket and the retaining heel of the gasket firmly bedded in its seat.

RUBBER RING JOINTS

Advantages:

- is easy and quick to assemble
- allows approximately 3 – 5° deflection depending on the size of the pipe
- permits movement of pipe caused by settlement, expansion, contraction or load.

Ensure that the gasket fits evenly around the whole circumference, removing any bulges that may prevent the proper entry of the spigot end of the pipe. A thin film of lubricant is applied to the inside surface of the gasket, which is in contact with the entering spigot (Figure 2.23(c)).

In addition, a thin film of lubricant may be applied to the outside surface of the entering spigot, for a distance equal to the depth of the collar. It is advisable to use only the lubricant recommended by the manufacturer as some lubricants are not compatible with the gasket material.

The spigot of the pipe being jointed must be aligned and entered into the socket carefully until it makes contact with the gasket. Final assembly of the joint is completed from this position. Joint assembly is completed by forcing the spigot end of the pipe past the gasket (Figure 2.23(d)), which is compressed, until the first painted stripe on the end of the pipe disappears and the second is approximately flush with the socket face (Figure 2.23 (e)). If the joint is difficult to assemble, the spigot should be removed and the gasket inspected for correct positioning.

Assembly of mechanical type rubber ring joints does not usually present problems. However, under certain conditions, such as wet trenches or large diameter pipes, additional leverage may be necessary.

On pipes from 100 mm to 200 mm diameter, a crowbar used as in Figure 2.24(a) is generally all that is required. It is important to have an assistant guide the spigot end of the pipe into the gasket to avoid malformation of the gasket.

Special pulling tools designed to ease assembly of these joints are also available. These tools consist of a forked bar, clamp or wire rope placed over the pipe, behind the collar, or actually on the collar, of the previously laid pipe and on the pipe being pulled (Figure 2.24(b)). A wire rope is wound around the spigot end of the pipe to be laid and is attached to the forked lever by a hook, or, where a clamp is used, it is joined by a bar. A lever is then moved in the direction of the arrow to pull the joint together.

In situations where a hydraulic trench-digging machine is being used for excavation purposes, the excavation bucket may be used to push home the pipe. Extreme care must be exercised when using this method to avoid possible damage to either pipe or gasket. Using a wooden block between the pipe and bucket helps to protect the pipe.

For pipes over 200 mm diameter, a rack and lever apparatus is used, which cranks the pipe slowly into the socket (Figure 2.24(c)).

FIG 2.23 (a) Rubber ring joint, (b) inserting the gasket, (c) lubricating the gasket, (d) initial entry of the spigot, (e) final assembly

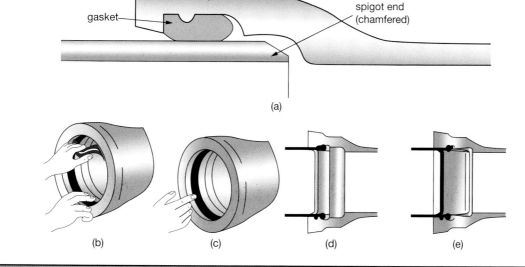

FIG 2.24 (a) The crowbar method of assembly, (b) the forked tool method of assembly, (c) rack and lever for large diameter pipes

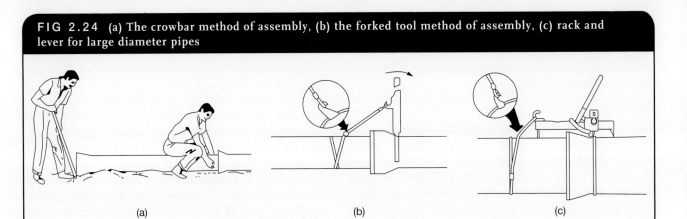

(a)   (b)   (c)

Flange joints

Ductile iron pipes designed to be jointed using flanges are manufactured in the same way as traditional pipes, except that the pipes are not cast with a socket on one end.

In order to attach the flanges to the pipe, they are first threaded in a lathe and an epoxy thread sealant is applied to the external thread on the pipe and the internal thread on the flange. The flanges are then machine-tightened until the thread protrudes past the face of the flange. The face of the flange and the protruding spigot are then machined off to provide a perfectly flat face to accommodate the gasket.

Flanged pipes must be laid in straight lines as there is no provision for deflection. The bolts used in the flanges must also be protected against external corrosion and selected to suit the soil conditions, for example 316 stainless steel for in-ground applications.

Ductile iron cutting

Pipes may require cutting, either to a specific length or because the spigot end has been damaged.

Various mechanical pipe cutters and/or power tools are available to cut ductile iron pipe. Eye protection and protective clothing should be worn when operating stationary or portable mechanical or power cutting equipment.

After a pipe has been cut in the field, the spigot requires chamfering to enable it to enter the collar easily without damaging the sealing ring and to remove any sharp edges. Refer to the manufacturer's instructions and recommendation as to the preferred method of cutting and chamfering.

Steel pipe

Steel pipe is generally used in the transmission section of a water supply system and to a lesser extent in distribution. Because steel has a much higher tensile strength than other materials used for the conveyance of water, pipes of very large diameters can be manufactured in comparatively light gauge material that still withstand very high pressures. Steel pipes are also cheaper, easier to construct and more easily handled and transported than ductile iron pipe.

Steel pipes were originally manufactured by rolling steel sheets to the required diameter either in one or more pieces, the longitudinal joints then being lapped and riveted. Circumferential joints on the ends of the lengths were usually fitted with flanges and bolted together. This method of manufacture has now been replaced by welded joints, although some riveted pipes are still available for special applications.

Steel pipe linings and coatings

Steel pipes are protected against internal corrosion by cement linings in a similar manner to ductile iron. External corrosion is a major disadvantage of laying steel pipes underground as steel, unlike ductile iron, has no inbuilt resistance to corrosion.

To protect steel pipes against the attack of aggressive soils, elaborate precautions are necessary. Careful soil analysis along the route of the proposed pipeline is carried out prior to specifying the type of coating to be used.

Steel pipe jointing

Steel pipes are usually manufactured with plain ends, which are designed primarily for welding. However, pipes are also produced with integral collars designed for use with rubber rings. Special rubber ring couplings and bolted flanges are also used.

Glass-reinforced plastic (GRP) pipe

NOTE: There are two types of GRP-pipe:
1. centrifugally cast (CC – GRP)
2. filament wound.
The text below describes text CC – GRP.

GRP pipe is now becoming increasingly accepted by Australian water supply authorities as a substitute for the previously common (now banned) asbestos- or fibre-reinforced cement pipe. GRP pipes are manufactured from chopped glass fibre roving, polyester resin and sand. They are centrifugally cast within an external mould and the walls are built up from the exterior surface to the interior surface. This is achieved by rotating the mould at a relatively slow speed and feeding the material in from one end at controlled rates. When all the material has been positioned inside the mould, the rotational speed is increased. The centrifugal force then compresses the composition against the inside of the mould, squeezing out any entrapped air and causing the materials to compact. Hot air is then passed through the spinning mould to accelerate the curing process.

The pipes produced have excellent corrosion-resistant properties and can be manufactured in large diameters, are lightweight and available in 6 m lengths. Diameters usually range from 150 to 1200 mm but are available up to 2900 mm for special applications.

Because of its manufacturing process, CC-GRP can withstand the compressive forces present during jacking operations. It can therefore also be used in trenchless installations.

GRP jointing

GRP pipes utilise the insertion-type jointing system, which consists of a coupling or sleeve with a rubber sealing membrane that is an integral part of the coupling or socket end of the pipe (Figures 2.25(a) and (b)). The rubber membrane covers the full width of the coupling and is attached to the pipe by simply pressing the sleeve over the end of the pipe after first lubricating both the inside of the coupling and the spigot end of the pipe. (GRP pipes are generally supplied with couplings fitted.)

GRP cutting

Because this pipe has uniform outside diameter, the pipe may be cut and joined at any point along its length. It is recommended that a water-fed abrasive disc cutter be used for this purpose. The same machine can be used to apply the 4 mm to 8 mm chamfer to the cut end to assist the entry of the pipe into the coupling.

GRP fittings

A range of tees and bends are available, however, the traditional spigot end ductile iron fittings may also be used on GRP pipe.

GRP transportation

The security of packaging depends on mode of transport (i.e. road, rail, sea, etc). When pipes of differing diameters are transported, 'nesting' (i.e. smaller diameter pipes inside larger diameter pipes) would normally be adopted (Figure 2.26).

FIG 2.25(b)   Large GRP pipe installation

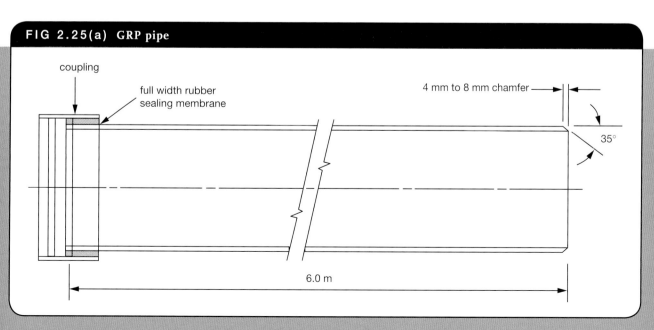

FIG 2.25(a)   GRP pipe

coupling

full width rubber sealing membrane

4 mm to 8 mm chamfer

35°

6.0 m

GRP storage

When GRP pipes are to be stored, the steel packing bands should be cut immediately after delivery and pipes restacked on wooden supports as shown in Figure 2.27.

Polyvinyl chloride (PVC) pipe

Because of the light weight and ease of handling, PVC pipes and fittings are being used increasingly for distribution pipelines throughout Australia, especially in country and remote areas.

They are manufactured by forcing a heated dry mixture of the basic ingredients through a set of extrusion dies to produce pipes of the required diameter and wall thickness. Pipes are available in diameters from 15 mm to 600 mm and in standard lengths of 6 m.

For water mains, PVC piping is available as PVC-U for applications up to PN18 and up to 300 mm in size. The addition of additives to make it more impact resistant result in the modified products, PVC-M for applications up to PN20 and 300 mm in size as well as PVC-O for applications up to PN25 and 600 mm in diameter.

Pipes are available with two in-built jointing methods and also plain-ended for specific applications.

PVC jointing

Joints used on pipes of this material are either the push-in rubber ring type or solvent weld joints and fittings are available for use with both jointing methods.

The rubber ring joint (Figure 2.28(a)) has the advantage of being able to be dismantled and remade if necessary. The solvent cement in the solvent weld joint has the effect of fusing the external surface of the pipe to the internal surface of the fitting, forming a permanent homogeneous bond between the pipe and the fitting (Figure 2.28(b)).

When laying pipelines with rubber ring joints, it is preferable to make the connections between pipes in the trench to avoid the risk of 'pulling' the joints when

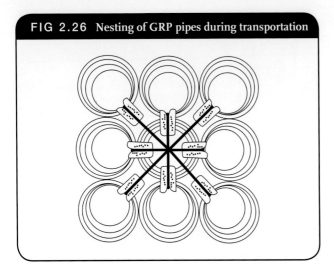

FIG 2.26 Nesting of GRP pipes during transportation

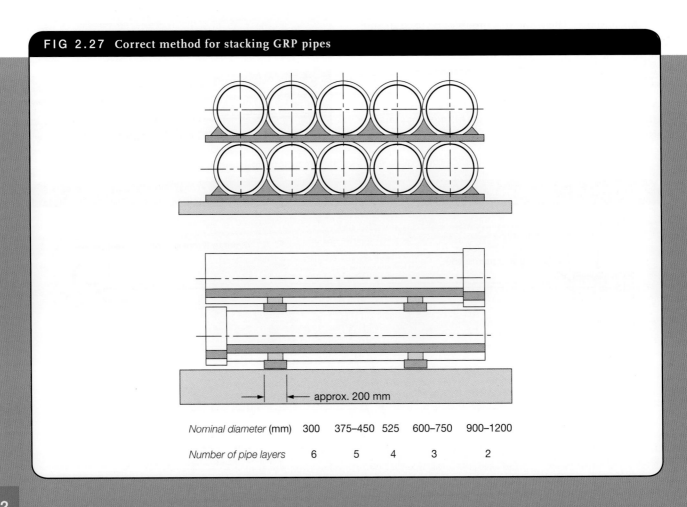

FIG 2.27 Correct method for stacking GRP pipes

approx. 200 mm

Nominal diameter (mm)	300	375–450	525	600–750	900–1200
Number of pipe layers	6	5	4	3	2

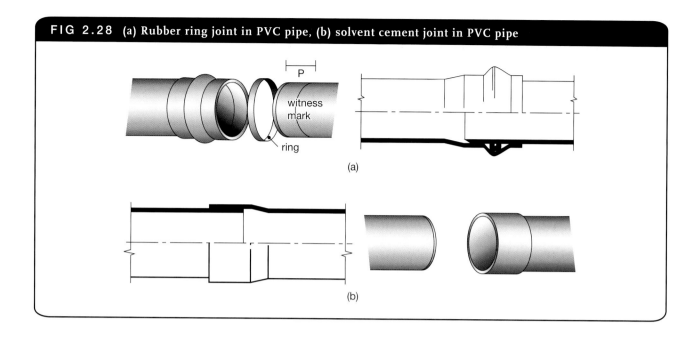

FIG 2.28 (a) Rubber ring joint in PVC pipe, (b) solvent cement joint in PVC pipe

(a)

(b)

handling the assembled line. Where the joints have been made outside the trench, every joint should be inspected after final positioning.

To assemble the joint, wipe out the ring groove in the socket to enable the rubber ring to make a clean, smooth contact. Similarly, clean the spigot end of the pipe back to the witness mark and ensure that the chamfer on the spigot is free from burrs. Insert the ring in the collar by forming the ring into a heart shape, which reduces its outside diameter. The ring should be clean and dry and is inserted with the thickest section to the back of the collar.

The spigot end of the pipe should be lubricated up to the witness mark, using the lubricant supplied. Oil or grease should not be used with rubber rings, as these substances will attack the ring and cause it to perish. If the correct lubricant is unavailable, a soap-and-water solution may be used as a substitute.

Align the pipes in the trench and push home, ensuring that the previously laid socket is firmly entrenched. Small diameter pipes are easily jointed by hand. However, larger diameter pipes may require additional force using a crowbar and a wooden block as in Figure 2.29(a).

PVC cutting and end preparation

The cutting of this material does not present problems and all PVC pipes may be cut by hand using a handsaw with fine teeth. To ensure the correct engagement of pipes into the fittings used, it is imperative that the ends of cut pipes are square. As this can be quite difficult to achieve, especially on large diameter pipes, a mitre box should be used whenever field cutting this material. Pipes can also be cut using a drop saw fitted with a fine-tooth blade and special cutting tools designed for cutting PVC.

End preparation prior to jointing with either rubber rings or solvent cement consists of chamfering the spigot end of the pipe. In the case of solvent weld joints

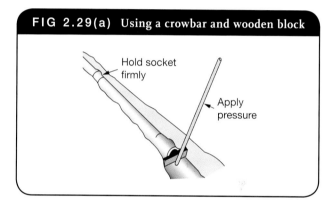

FIG 2.29(a)   Using a crowbar and wooden block

Hold socket firmly

Apply pressure

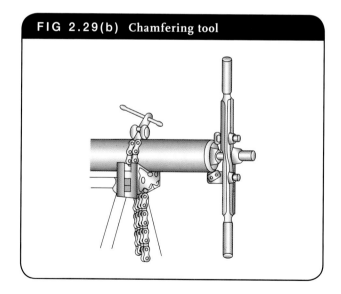

FIG 2.29(b)   Chamfering tool

this chamfer can be made using a coarse file or rasp. When using rubber ring joints, the chamfering of spigot ends needs to be more accurate so that the entry of the spigot into the collar or fitting is a smooth operation that does not deform or misplace the ring.

There are numerous designs of chamfering or bevelling tools available. A typical example is shown in Figure 2.29(b).

Polyethylene

Polyethylene is becoming more popular due to its ease of installation, flexibility, long lengths and ability to be installed with trenchless technology. The bore is very smooth, making friction losses less than other materials. It also has high-impact resistance and a predicted long service life of over 100 years. Its suitability for the medium it carries is usually identified by a coloured stripe on black pipe. Drinking water is identified by a blue stripe or blue coloured pipe (NZ) and recycled water with a purple (lilac) stripe. Non-potable water is simply plain black pipe as with irrigation piping. Figure 2.30(a) demonstrates the flexibility of the complete system.

Polythylene jointing

Mechanical compression fittings are commonly used for pipes up to 110 mm. The fusion method of jointing (Figure 2.30(b)) is used for 90 mm and above (although smaller pipes can also be fused) ensuring a high strength joint. PE flange stubs for use with mechanical jointing can be welded onto pipes up to 160 mm in diameter and flange stubs for fusion welding can be used up to 450 mm. Operators need to be properly trained in electrofusion and butt fusion welding techniques before attempting these methods of jointing. Refer to the manufacturer's instructions as to the correct installation procedures and methods of jointing.

HANDLING AND STORAGE OF MATERIALS

The materials used in water distribution systems are necessarily of robust construction, enabling them to withstand extreme internal pressures and external loads. However, so that they can withstand attack from aggressive soils and the water they convey, surfaces that come into contact with these substances must be protected with some type of protective coating. Metallic pipes are particularly

FIG 2.30(b) Butt fusion welding in the field

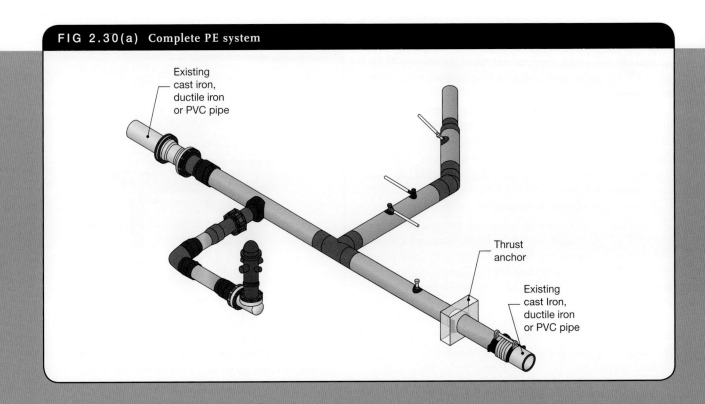

FIG 2.30(a) Complete PE system

Existing cast iron, ductile iron or PVC pipe

Thrust anchor

Existing cast Iron, ductile iron or PVC pipe

susceptible to this type of attack and although the pipes themselves are able to withstand rough treatment, the linings and coatings will not.

Non-metallic materials are easily damaged if not handled correctly. Excessive heat may distort and ultimately destroy the pipe and fittings if left for prolonged periods exposed to ultraviolet light. The storage area needs to be clean, flat, secure, and protected from the weather and surrounding environment to maintain the integrity of the materials.

Unloading

NOTE: The following information is meant as a guide only. You should refer to current safety regulations for the correct procedures for both loading and unloading.

The procedures chosen will depend upon the site conditions, what is being unloaded and the type of vehicle it is being unloaded from. As with any procedure, a safe method of working must be identified and documented.

Before unloading any pipe materials, personnel should be aware of the weight of the material to be lifted and the difficulties presented by its length. Details of the mass of the different pipe materials and diameters are readily obtainable from the manufacturers.

Units or pallets

Small diameter pipes of all materials are sometimes delivered on site packed in units or on pallets and are usually unloaded using mechanical means (Figure 2.31(a)). Before unloading pipes packed in this way, ensure that all personnel are correctly allocated mechanical equipment such as cranes and forks of adequate capacity, and that relevant regulations are observed including training and licensing in operation of equipment.

Single pipes

The method adopted for unloading single pipes depends on the diameter and weight of the pipes and the height to which they are stacked on the delivering vehicle. They should only be unloaded by hand where they are easily accessible and do not pose a hazard in manual handling. Even though a pipe may be relatively light, its length may make it difficult to unload manually. Each situation should

FIG 2.31(b) Unloading large pipes with ropes and slings

(a)

(b)

C hook

rubber covered end

(c)

(d)

be assessed and controlled for the hazards it presents, and appropriate measures employed to ensure safety.

Pipes with large diameters should be unloaded individually using a crane and rope slings as in Figure 2.31(b).

Unloading along the trench

Pipes should only be unloaded along the trench where unloading will not create a hazard and where they are to be laid immediately, as they are susceptible to damage by either earth-moving or trenching machinery. Also the linings and coatings may be damaged or contaminated if the pipes are covered with material excavated from the trench. If pipes are to be laid along the trench line, the following points should be observed:

1   Unload the pipes as near to the trench as possible to avoid excessive handling.

2   Unload pipes on the opposite side of the trench to the spoil. This enables the pipe layers to roll the pipe up to the edge of the excavation without difficulty.

3   Ensure that all necessary fittings are laid in their correct positions in accordance with the plans.

FIG 2.31(a) Unloading pipes using a forklift

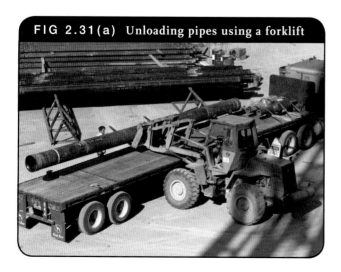

Unloading fittings

Fittings, like pipes, should be handled with care so that they are not damaged by being dropped or thrown against each other. If the fittings are being unloaded by mechanical means, the hooks or slings used should be padded, especially where they come in contact with open ends.

Large diameter ductile iron fittings are usually fitted with lifting lugs, which are positioned at the point of balance of the fitting.

Stacking

NOTE: The following information is meant as a guide only. You should refer to current safety regulations for the correct procedures for stacking pipes and fittings. As with any procedure, a safe method of working must be identified and documented.

The following precautions should be taken if pipes are to be stacked on site prior to use.

Ensure that the area on which the pipes are to be stacked is level. Several 400 mm and 75 mm timber battens are required with chocks already fitted at one end. Place two of the battens on the ground approximately 2.5 m apart and parallel to each other. Build the stack in a similar manner to the way the pipes are stacked on the delivery vehicle. When the first row is completed, fix a chock to the cleat on the end nearest the delivery vehicle and place the next two battens directly above the first two. Continue the stack until the desired number of rows is reached.

Stacking PVC pipes

Pipes and fittings manufactured from PVC may be unloaded and stacked in a similar manner to other materials used for water supply systems. However, because the material is easily distorted when subjected to excessive pressure or heat, the number of supporting battens required between the rows is greatly increased. PVC pipes are usually supplied in 6 m lengths, although longer lengths are also available for special applications. When stacking 6 m lengths, five 75 mm battens are required, positioned so that the distance between them does not exceed 1 m.

The pipes should be laid on the battens so that a spigot end rests against a socket end on alternate pipes. This ensures that the pipes lie parallel to each other in the stack. Side supports are also required and these should be spaced not more than 2 m apart.

Stored PVC pipes should always be covered to prevent direct contact with the sun's rays, as uneven expansion resulting from direct sunlight on one side of the pipe will cause bowing or twisting. High temperatures tend to soften PVC which is then easily distorted by the weight of the pipes in the stack.

Because PVC pipes are easily handled, there is a tendency to stack them too high, which increases the risk of damaging the pipes in the lower part of the stack.

When stacking pipes of different classes, always place the pipes with the greatest wall thickness at the bottom of the stack.

PVC pipes should not be laid along the side of the trench unless they are to be installed and backfilled immediately.

Fittings should also be stored away from sunlight until they are required.

Unloading and stacking polyethylene (PE)

Procedures for handling PE pipe are similar to PVC. The main difference is that PE comes not only in straight lengths but also in coils and as with any material, care in handling needs to be exercised so as not to distort or damage the pipe. The storage area should be relatively flat, smooth and free from rocks or material that could damage the pipe. Coils should be unloaded mechanically and never dragged or pushed. The pipe and fittings need to be visually inspected before storing to ensure that they are free from defects that would affect their performance and longevity.

TAPPING THE MAIN

Metallic mains may be tapped directly to the main by drilling and threading the pipe wall and screwing the main tap directly into it or a tapping band can be used. Non-metallic mains require a tapping band (Figure 2.32(a)) that wraps around the pipe and houses the main tap. The exception is polyethylene, where an electrofusion tapping saddle (Figure 2.32(b)) should be used on all new installations. Tapping bands may be used for PE under certain circumstances but only where it is impractical to use a fusion saddle.

Main taps are usually 20 mm or 25 mm and installed singularly or as a pair connected to a breeching tee for water services up to 65 mm. Where more than one tapping is required refer to the manufacturer for the minimum distance between tappings. Water services 80 mm and over have a tee and sluice valve either cut into or saddled onto the watermain. Refer to Chapter 3 for more detail.

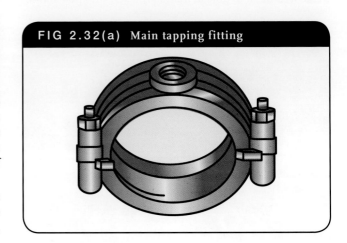

FIG 2.32(a) Main tapping fitting

FIG 2.32(b) Electrofusion saddle tapping

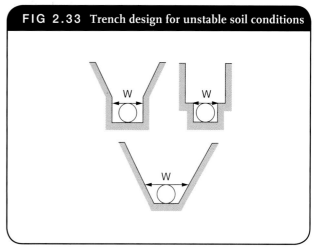

FIG 2.33 Trench design for unstable soil conditions

EXCAVATION OF TRENCHES

Trenches should be excavated in accordance with plans and specifications. They should be as narrow as practicable at a point level with the top of the pipe, and in straight trenches the bed of the trench should be not less than 200 mm wider than the pipe to be laid. The 100 mm allowance on each side of the pipe is required to provide a working space for the pipelaying crew.

Additional width in the bed of the trench is required where the pipeline is to be laid around a curve or in situations where the trench is excessively deep.

The ground in which trenches are to be excavated is generally classified as either 'stable' or 'unstable'. The factors that affect the classification of a trench are:

1 soil type

2 amount of moisture present

3 trench width, which is controlled by the diameter of the pipe

4 trench depth

5 method of excavation.

Stable conditions are those where the trench walls remain solid after excavation and do not show signs of collapse or cave in.

Unstable conditions are those where the trench walls tend to collapse and cave in during or after excavation. Under these conditions, in open or unrestricted areas, the top of the trench may be widened or battered until the stability of the surrounding soil is achieved. A smaller trench should then be excavated in the bottom of the original trench to accommodate the pipe (Figure 2.33).

In areas where the width of the trench is restricted, for example, in a public street, between existing buildings or where the trench is to be dug in dry running sand, it may be necessary to support the sides of the trench by shoring the walls with either timber supports or other suitable materials (refer to *Plumbing Services: Drainage* for further information on trench support).

When these precautions are necessary, it is advisable to contact the appropriate authority so that the correct method and materials are used to comply with local safety regulations.

Preparing the trench

When preparing a trench for pressure pipelines, the bed of the trench should be excavated to form a level platform on which the pipes can be laid with their barrels supported for the entire length of the pipe. In good working conditions, such as sandy or loamy soils, the bed of the trench can be excavated to form a level base without the addition of a bedding material on which to lay the pipes. In shale or rocky ground, however, it may be necessary to import suitable bedding material on which to lay the pipeline. The two most commonly used materials are coarse river sand or blue metal. However, the final choice of material depends on the pipeline material and the conditions in the trench at the time of laying. If conditions in the trench are wet, with water running down the excavation, then sand is unsuitable as a bedding material because of the scouring effect of the water. In this situation, fine metal screenings are the most appropriate material.

When a trench is being excavated to receive a pipeline, it is recommended to dig 50 mm deep recesses to accommodate joining couplings and fittings. These recesses ensure that the pipes are supported on their barrels and also reduce the possibility of soil entering the pipes and fittings, which could damage the rubber rings and make jointing of pipes in the trench extremely difficult (Figure 2.34(a)).

An alternate method that can be employed in situations where imported bedding is used is to support the pipes on mounds placed 150 mm back from the coupling positions, with the mounds running right across the trench. The pipes are then supported on these mounds and fine backfill placed and tamped well under the pipes after completion of the joints.

As a guide, the amount of fine backfill or imported bedding used under pipes should be such that, after levelling, the highest projections in the trench bottom are covered by at least 75 mm. In rock conditions this covering should be increased to 150 mm (Figure 2.34(b)).

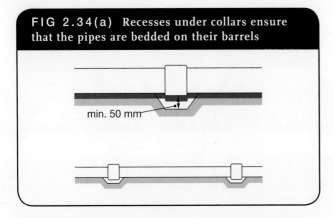

FIG 2.34(a) Recesses under collars ensure that the pipes are bedded on their barrels

min. 50 mm

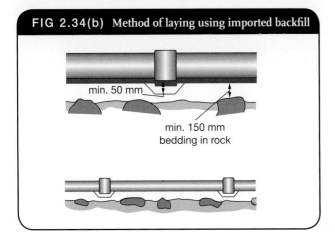

FIG 2.34(b) Method of laying using imported backfill

min. 50 mm

min. 150 mm
bedding in rock

Backfilling the trench

Initial backfilling

The main aim in initial backfilling of a trench is to provide support for the pipeline, so that when backfilling is completed the load of fill in the trench, plus any live loads, does not cause the pipe to settle. If settlement occurs, the grade will be upset and the pipe may be damaged, resulting in costly and time-consuming repairs.

Backfilling material used for this purpose should be free from stones, rock or clay. If this type of material is not available on site, a suitable overlay material should be imported. Loamy earth or sand is ideal for this purpose. The initial backfill should be placed around the pipes by hand in layers not exceeding 100 mm deep. Each layer should be carefully tamped around the pipe, with special attention being paid to the underside of the pipe. The initial backfilling procedure of placing 100 mm of fill and tamping should be continued until the effective cover of fill over the pipe is at least 300 mm.

When carrying out initial backfilling, it is good pipelaying practice to leave pipe joints exposed until after pressure testing has been completed to ensure that faulty joints may be easily identified. It is generally stated in pipelaying specifications that these field joints be left uncovered so that inspections by the controlling authority can be carried out while the pipeline is under pressure test conditions.

To complete initial backfilling, the selected backfill or overlay should be hand shovelled over the exposed joints so that a continuous depth of 300 mm is maintained over the entire length of the pipeline.

Final backfilling

After the pipeline has been pressure-tested and the selected spoil or backfilling material placed over the exposed joints, final backfilling can be completed to ground level. The material used is usually spoil that has been previously excavated from the trench and can be placed either by hand or by mechanical means.

Even though the pipeline is protected by 300 mm of selected fill, care should be exercised when completing the backfilling operation to ensure that any large rocks are removed from the final backfill as they may work their way down through the fill and eventually come in contact with the pipe. This is especially so in situations where the excavation is subjected to live loads, as would be the case if the trench were located in a public street.

ANCILLARY EQUIPMENT FOR WATER DISTRIBUTION SYSTEMS

Repairs

During the life of a water distribution system it often becomes necessary to carry out repairs or alterations to a previously installed pipeline. These repairs and alterations are carried out in difficult situations, generally in unsatisfactory working conditions and usually at short notice.

To alleviate the difficulties encountered in this type of work, a series of repair fittings are available that have been designed for a specific type of breakdown or for general use in an emergency.

Repair coupling

This fitting (Figure 2.35) consists of a stainless steel jacket and a rubber sealing sleeve that can be bolted around a pipe to repair a circumferential fracture. This fitting is designed for use on straight lengths of pipe and is unsuitable for use in situations where the pipe is to be deflected.

FIG 2.35 Repair coupling

Flanged offtake clamp

This fitting (Figure 2.36) is attached in a similar manner to the repair coupling and is used where a flanged outlet is required. It is ideal for providing additional branch lines, from an existing pipeline. The control valve for the branch is bolted directly on to the flange.

Socket repair joint

As the name implies this fitting is used for repairing leaking or damaged sockets or couplings. It is designed to completely seal the socket and is attached to the barrel of the pipe on both sides using a rubber gasket (Figure 2.37).

FIG 2.36   Flanged offtake clamp

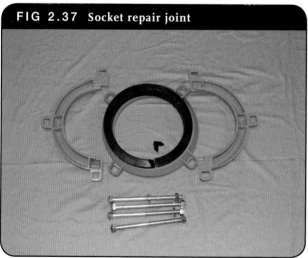

FIG 2.37   Socket repair joint

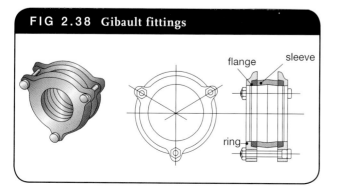

FIG 2.38   Gibault fittings

flange   sleeve

ring

Gibault fittings

Gibault fittings (Figure 2.38) have the distinct advantage that they may be fitted to pipes that are not correctly aligned or where a fracture occurs in a pipeline that has been laid round a curve. They are available in different configurations to suit a variety of materials and even join materials of unequal size or different composition.

Thrust blocks

Distribution systems are designed to convey water at extremely high pressures and velocities. These internal stresses and their effect on the system are usually considered when the pipeline is being designed. The route of the pipe, types of material used, positioning of the valves and ancillary equipment are carefully planned so that the internal stresses are distributed as evenly as possible throughout the entire system. In practice however, because of the restricted space in which these pipes are generally laid, controls and sharp changes of direction are unavoidable and these restrictions to the flow of water concentrate the internal stresses in these areas. For this reason thrust blocks are installed on the pipeline in positions where increased pressure resulting from the closing of a valve or an acute change of direction is likely to occur (Figure 2.39).

For thrust blocks to be effective, the size of the bearing surface of the block is carefully calculated and the bearing value of the surrounding ground is also of great importance. Pipelines that are jointed by 'push-in' or mechanical joints are more likely to suffer damage from internal stress or thrust than those with welded or flange joints, as the pressure inside the pipework can push the ringed fitting/s off the pipe. Thrust blocks should be designed by a structural engineer with suitable experience.

Air valves

Water has the ability to absorb air and other gases into solution. This air is periodically released from the water by changes in temperature, movement and compression, and under normal circumstances collects at high points in the distribution system.

As the air accumulates, it reduces the pipe diameter, which reduces the discharge potential or increases the friction head. For this reason pressure pipelines should be laid evenly to grade and changes in vertical altitude should be kept to a minimum.

To enable this air to be released from the line, air valves should be positioned at the high points in the line. These high points, or 'peaks' as they are known, are located by reference to the hydraulic gradient and not to a horizontal line.

Air valves are required more often in the transmission system than in the distribution system, as it is possible to release air from the distribution pipework through fire hydrants and the individual service connections. They occur more often in the distribution systems of sparsely populated areas.

FIG 2.39 Details of thrust blocks: (a) bend in horizontal plane, (b) bend in vertical plane, (c) reducer anchorage, (d) tee anchorage, (e) valve anchorage, (f) closed end and hydrant

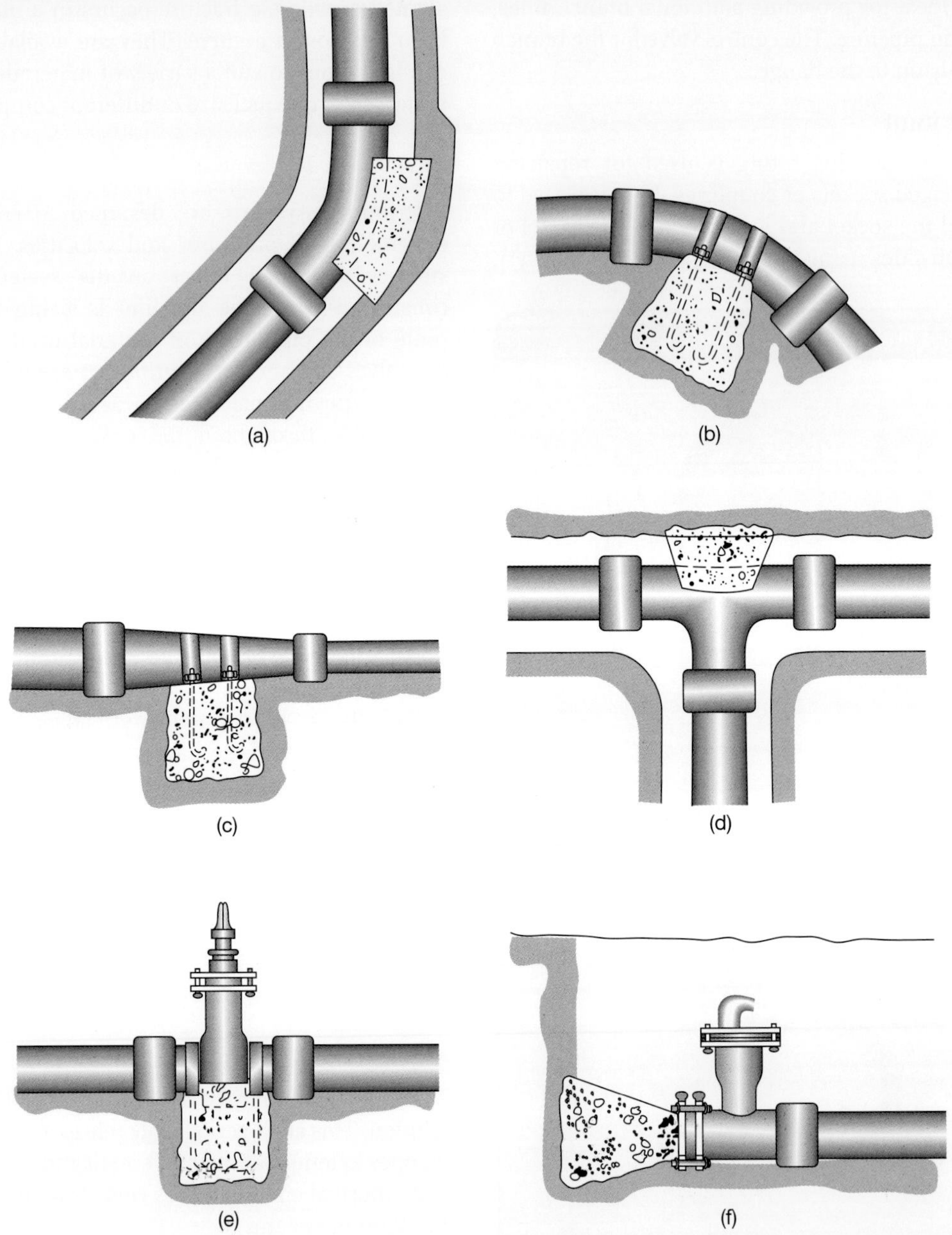

Both single and double air valves are available. The single air valve is used to release small quantities of air that may accumulate at high points in a charged water main (Figure 2.40). Double air valves are fitted with both small and large orifices in separate chambers and are designed to release small quantities of air in a similar manner to the single air valve. They also release or admit large volumes of air when a pipeline is drained or filled rapidly (Figure 2.41).

Identification of boxes and covers

There are numerous standard boxes and covers used on distribution systems. They are usually manufactured in cast iron, concrete, plastic or fibre cement, and because of their locations are required to be strong, durable and able to withstand extremely heavy loads.

In most instances they are used in an emergency and for this reason they need to be easy to locate. To assist in the rapid location of this ancillary equipment, reflective signs and symbols are usually attached to some permanent fixture such as a light pole or fence adjacent to the underground equipment. These signs indicate the approximate position and type of equipment located beneath the cover or box. Examples of some of the standard covers used and the states of origin are shown in Figure 2.42.

FIG 2.40 Single air valves release small quantities of air from a pipeline

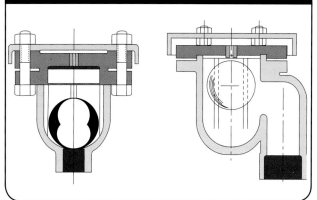

FIG 2.41 Double air valves serve a dual purpose as they are able to discharge both large and small quantities of air

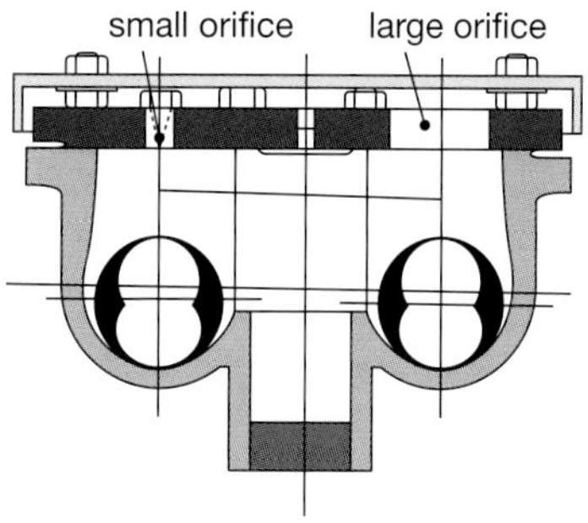

FIG 2.42 Standard covers

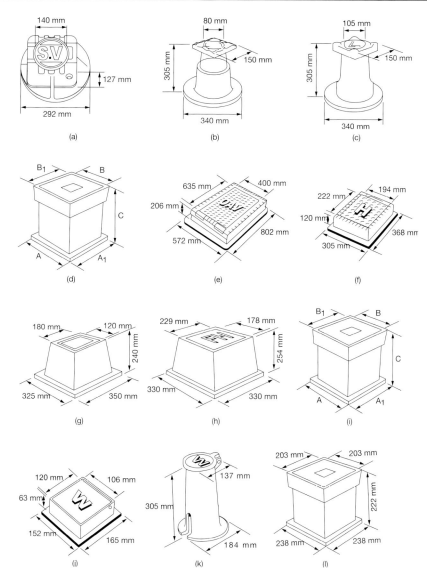

(a) Stop valve box (NSW, SA, NT) (b) Valve cover box (Vic.) (c) Stop valve box (Qld)
(d) Stop valve box (WA) (e) Double air valve box (f) Hydrant box (NSW, SA, NT) (g)
Fire plug cover (Vic.) (h) Hydrant box (Qld) (i) Hydrant box (WA) (j) Path box (NSW,
SA, NT) (k) Path box (Qld) (l) Path box (WA)

FOR STUDENT RESEARCH

AS/NZS 3500 Part 1: Water services, provides acceptable solutions so water systems can comply with the Plumbing Code of Australia and the New Zealand Building Code. AS/NZS 3500 cites many other standards relevant to the materials and their installation. Refer to these standards to further your understanding of water services and water piping systems. Also refer to the WSA 03-2011 Water Supply Code of Australia.

Visit the following websites for more information on water services, materials and environmental issues.

- www.livinggreener.gov.au/water/greywater
- http://rainharvesting.com.au/rainwater-knowledge-centre/how-to-create-a-complete-system
- www.vinidex.com.au/page/ductile_iron_pipe_saint_gobain_pam.html
- http://hobas.com.au/HOBAS-Water-Pipe.php
- www.promains.com.au/static/pdf/promains-grp-pipe-and-fittings-product-guide-may-2012.pdf
- www.wsaa.asn.au/Codes/Documents/Water.pdf

Australian Standards

AS/NZS 3500:1 Plumbing and drainage—Water services

WSA 03-2011 Water Supply Code of Australia

ON-SITE STORIES 2.1

SOMETIMES, EVEN WHEN YOU THINK YOU HAVE FOLLOWED ALL THE PROCESSES, YOU STILL END UP IN TROUBLE!

Bruce Paulsen, Plumbing *and Sustainability Teacher, North Sydney TAFE; co-author of Basic Skills*

Most plumbers know what to do when excavating in a public place. We all know that if you lodge an online enquiry with 'Dial Before You Dig' you will be given details of all the services in the area you wish to excavate. In fact, it's a legal requirement to do this whenever we excavate in public property.

So anyway, having done my searches, I knew that there was a 150 mm cast-iron water main in the street where I had to excavate for the sewer connection. Carefully positioning his machine, my excavator driver slowly began exposing the main. First, he had to move a large boulder that was sitting on the nature strip, so he reached out with his bucket and started pulling the rock across the ground. We watched in horror as the ground opened up and a torrent of water poured across the nature strip and down the kerb and gutter.

Well, Sydney Water came and turned off the water to the street, and while fighting off angry residents, we exposed the main. We found a 3 metre-long crack in the main which had opened up when the weight of the boulder had been removed.

Sometimes it's better just to stay in bed!

Water services and meters

INTRODUCTION

The 'service' is that part of the water supply system conveying water from the authority's main to the metering device, generally located inside the boundary of the consumer's property. The responsibility for laying the service pipe varies from state to state. In some areas, it is the responsibility of the water supply authority to tap the main and lay the service, while in other states opening of roads and the excavation of trenches ready to receive the service pipe is the plumber's responsibility.

WATER SERVICES ASSOCIATION OF AUSTRALIA (WSAA)

Water and sewage services in Australia are vital not only to people and households but also to industry and commercial enterprises and are provided by government-regulated water utilities. The Water Services Association of Australia (WSAA) is the peak industry body that brings together and supports this Australian urban water industry. Members provide water and sewage services to more than 16 million Australians. They also provide services to many of Australia's largest industries and commercial enterprises.

As the peak body representing the nation's urban water industry, WSAA acts on behalf of all members, providing a strong, national voice for the sector and taking a leading role in influencing urban water policy development. The WSAA regularly assesses and reports on the performance of the industry to help support members in their engagement with customers, stakeholders and the community. To implement the WSAA requirements, each state and territory also has supplementary documentation for their installation requirements.

Payment of fees

Before starting work on the laying of the service it is usually necessary to pay fees to both the water supply authority and local government authority. The fees payable to the water supply authority are intended to cover the tapping of the main, initial and subsequent inspections (which are required when carrying out all water supply work) and, in some instances, the supply of the service control valve and fittings.

The fees payable to the local government authority are intended to cover site inspections and restoration of road and footpath surfaces should these areas be damaged during the excavation and laying of the water service. In new subdivisions where roads are surfaced, and kerbing and guttering is provided, the underground mains are already laid, to prevent disturbance of the existing structures.

The opening of roads and footpaths is prohibited in some areas. Where these restrictions apply, a service conduit is supplied so that the water service may be laid without disturbing the road surface. The position of the conduit is usually marked on the kerb or by the use of coloured pegs, and notification signs are placed in prominent positions in the area (Figure 3.1).

FIG 3.1 Prominent signs are erected directly above conduits

LOCATING THE MAIN

Information about the position of water mains in relation to other underground services, such as gas mains, underground power cables and telephone cables, may be obtained from the relevant authority when fees are paid. In many cities, and especially on main roads, underground services are allocated fixed positions under footpaths and roadways. These positions are agreed upon by the various regulatory authorities. Details can be obtained by telephoning Dial Before You Dig (DBYD). Care must, of course, be exercised by the plumber during excavation.

NOTE: Prior to the commencement of any works, the contractor/licensed plumber is required to obtain the location of all services from Dial Before You Dig (DBYD) by telephoning 1100 or applying through selected outlets or online.

MATERIALS

Ductile iron cement-lined pipe (DICL)

Where DICL pipe is required for water main applications it will be detailed on the approved design drawings. Ductile iron pipe that is manufactured in accordance with the current AS/NZS 2280: 1999 uses a K(n) value to describe the class of pipe, for example, K9. The pipe uses a rubber ring jointing system and is internally lined with a cement mortar as specified in AS/NZS 2280. For example, SA Water has nominated class K9 for all reticulation applications with an operating pressure of less than PN16.

Polyethylene pipe

Polyethylene pipe has become one of the most widely used of all plastic systems within the water supply industry. Its advantages are its corrosion resistance, light weight and ease of installation. Long lengths are used for directional drilling and in country regions where long runs are required. A detectable underground warning tape is to be laid on top of the embedded pipe, which is a requirement of the Water Supply Construction Manual (WSCM).

Jointing methods for polyethylene pipe include the following:

- butt welding, which provides homogenous joints and is essential to ensure adequate weld strength; must be carried out by a suitably skilled and /or certified operator

- mechanical compression

- electrofusion welding, using electrofusion couplings requiring a controlled electrical input from a welding machine.

Other important characteristics of polyethylene are its high flow capacity, high impact strength and weather resistance. The pipe is stabilised against ultraviolet light degradation by the inclusion of carbon black in the raw material, making it suitable for situations where it will be exposed to direct sunlight.

PVC-U

PVC-U (unplasticised polyvinyl chloride) accounts for a large proportion of plastic pipe installations. This is due to its excellent chemical resistance and broad range of operating pressures, ease of handling and installation, material strength and outstanding flow characteristics. PVC-U is manufactured in Australia, conforming to AS/NZS 1477: 1999.

PVC-O

PVC-O pipes are used for medium- and high-pressure water pipe systems for potable water supply, firefighting mains, recycled water, irrigation and pumping systems. It is suitable for these purposes because of the molecular orientation manufacturing process, which strengthens the pipe to prevent cracking due to scratches. PVC-O pipes are able to absorb excessive pressure caused by water hammer. The pipes are recyclable as they can be ground up and reprocessed for use in the manufacture of other plastic products.

Existing service connections (steel and cast iron mains)

When a service control valve is to be connected to an existing steel or cast iron mains, the main is usually drilled and tapped by the water supply authority, or selected subcontractors on behalf of the authority, and the service control valve is attached. In the majority of states the most common tapping size is 25 mm BSP thread. When the service diameter exceeds 25 mm, two drillings are required, connected by a breeching piece to the service pipe (Figure 3.2).

When a service is required that exceeds 65 mm in diameter, the main is cut and a special flanged tee is used

FIG 3.2 Breeching piece

to which the control valve is attached. This work is carried out by the authority in the majority of cases.

Gibault elongated joints

These fittings are designed for use on repairs to steel or cast iron mains. They are similar to the standard repair joint except that the centre section is elongated to accommodate the tapped boss. They provide the added advantage of being able to receive a larger service than would be possible by directly tapping the main (Figure 3.3).

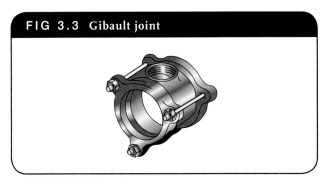

FIG 3.3 Gibault joint

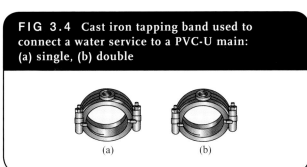

FIG 3.4 Cast iron tapping band used to connect a water service to a PVC-U main: (a) single, (b) double

(a)　　　　(b)

Service connections (PVC-U mains)

When a service is to be connected to a PVC-U main, a cast iron tapping band is used (Figure 3.4). This is manufactured in two halves, fitted with neoprene or rubber seals, and bolted together. The connecting bosses are threaded to receive the service control valve and may be attached to the main while it is under pressure using a specially adapted drilling machine.

EXCAVATING THE MAIN

Every state and territory will have a Road Management Act that outlines the responsibilities for conducting works on roads and traffic management, and executing a duty of care. An example of this is the Victorian *Road Management Act 2004* which contains the 'Code of Practice, Worksite Safety—Traffic Management'.

It is always the contractor's responsibility to ensure that the safety and welfare of both plumbers and the public are considered. This is especially the case where excavations are carried out on private property or in a public street. It is also the plumber's responsibility to supply and erect adequate warning signs, barricades and, where necessary, covering for all excavations. Where an excavation is to be left open overnight, adequate lighting must be provided.

Where an excavation is to be undertaken on a public roadway, special care should be taken to ensure the safety of both the workers and the traffic. Figures 3.5 and 3.6 show examples of the minimum safety requirements when excavating a trench in a public thoroughfare.

Road markings should be positioned to prevent confusion. Approaching traffic is to be guided gradually to either side of excavations in the centre of the road, or around those at the edge of the road. The positioning of a protective vehicle is most important, as it provides a last line of defence for

FIG 3.5 An example of the minimum safety requirements for: (a) a trench in the centre of a road, (b) a trench near the edge of a road

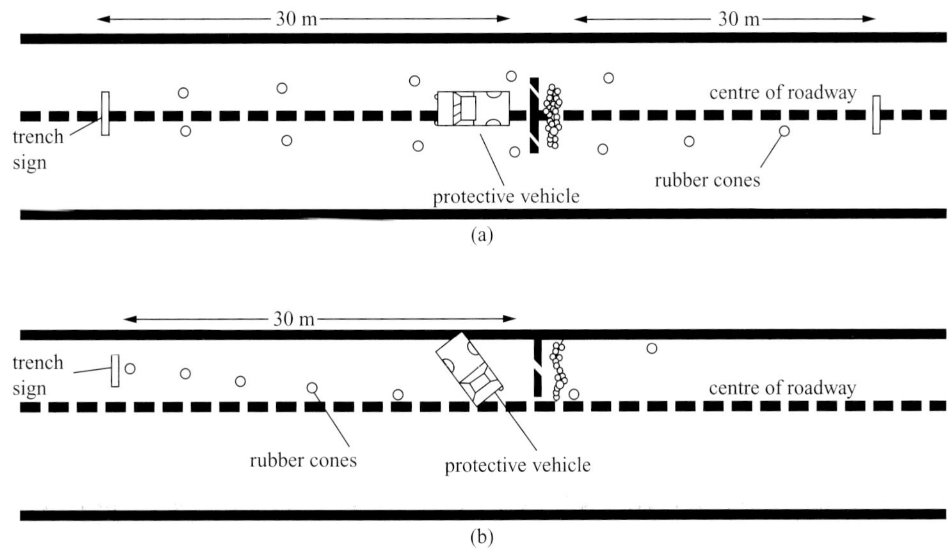

30 m　　　　30 m

trench sign

centre of roadway

rubber cones

protective vehicle

(a)

30 m

trench sign

centre of roadway

rubber cones　　protective vehicle

(b)

FIG 3.6 An example of traffic control

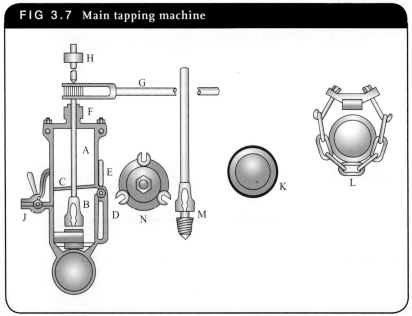

FIG 3.7 Main tapping machine

The main components of a mains tapping machine are:
A. chamber
B. chamber
C. watertight flap
D. bypass
E. valve
F. packing gland
G. ratchet handle
H. supporting frame
J. drain valve
K. sealing gasket
L. tensioning chains
M. tool spindle and holder
N. retaining flange.

workers should the driver of a vehicle fail to observe the warning signs.

DRILLING AND TAPPING CAST IRON MAINS (UNDER PRESSURE)

Drilling and tapping cast iron mains under pressure has distinct advantages over the drilling and tapping of mains with the water turned off, even though the equipment required is more complicated to use and takes a little longer. When the pressure in the main is reduced, as is the case when the water is turned off, small, previously unnoticed leaks become entry points for dirty or polluted water. When the main is turned off, ball-type hydrant valves in which the internal pressure of the water holds the ball seal in position drop open, allowing entry of polluted water into the main. Air may also enter the main at this time, causing inconvenience to consumers and accelerating internal corrosion of the main.

Figure 3.7 shows a mains-tapping machine used for drilling and tapping mains under pressure. The machine consists of two chambers, A and B, separated by a watertight flap, C. The two chambers are also fitted with a bypass, D, controlled by a valve, E, which enables the pressure in both chambers to be equalised.

Positioning the service control valve

There are four factors that influence the position of the service control valve:

- the position of the main and the material from which it is constructed
- the type of service control valve used (must be approved by the water control authority)
- the position of the service conduit (if provided)
- the position of the meter (the meter should be located at 90 degrees to the main).

PROCEDURE FOR TAPPING MAINS UNDER PRESSURE

1. Locate and excavate the main, making sure that there is ample room in the excavation in relation to its depth.

2. Mark the desired position on the main using a centre punch or diamond point chisel (Figure 3.8(a)).

3. Place the sealing gasket in position, with the starting mark centralised (Figure 3.8(b)).

4. Attach the machine to the main using the tensioning chains, taking care that the machine is located squarely on the sealing gasket (Figure 3.8(c)).

5. Place the tapping tool in the tool holder and fix the retaining flange and spindle to the machine (Figure 3.8(d)).

6. Proceed to drill and tap the hole in the main (Figure 3.8(e)).

7. After the hole has been drilled and tapped to the required depth, remove the spindle and replace the tapping tool with the main control valve or ferrule (Figure 3.8(f)).

8. Replace the retaining flange and spindle to which the main control tap is attached. Screw the control tap into the main (Figure 3.8(g)).

9. Remove the machine from the main and give the control tap a final tighten (Figure 3.8(h)).

Most water authorities will not permit the tapping of mains within a specified distance of a collar or fitting: it is vital to check requirements with the authority prior to tapping.

DRILLING AND TAPPING CAST IRON MAINS (WATER TURNED OFF)

The equipment required for drilling and tapping mains when the water is turned off is much simpler than that required for under-pressure tapping. However, the advantages of tapping under pressure are lost when the water is turned off.

SEPARATION OF SERVICES

Separation from above and below ground electrical supply cables, consumer gas pipes, water mains and underground obstructions are clearly defined in AS/NZS 3500.1.

FIG 3.8 Main tapping

LAYING THE SERVICE

NOTE: The service should be laid in a straight line from the main to the meter position and always at 90 degrees to the main. This positioning is standardised so that the service pipe may be located easily at a later date.

As previously mentioned, service conduits are sometimes provided where the service pipe is required to pass under roads. This is the exception rather than the rule and usually occurs only in newly built areas where roads are sealed and kerbs and gutters are already installed. In situations where a service is to be laid across a road and a service conduit is not provided, an alternative method must be found to locate the service pipe in the desired position.

There are four accepted methods of installing service pipes under roads. They are:

1   excavating

2   driving

3   jetting

4   boring.

Excavating

NOTE: Prior to the commencement of any works, the contractor/licensed plumber is required to obtain the location of all services from Dial Before You Dig (DBYD) by telephoning 1100 or applying through selected outlets or online.

Excavating is only suitable where an unsealed road has to be crossed or it is known that the composition of the ground through which the service has to pass contains a high percentage of rock. Main roads authorities and local councils seldom permit the excavation of tar-sealed or concrete roads for the laying of services, unless other methods have been tried and have proven unsatisfactory. If permission has been granted by an authority to open a sealed or unsealed road, the excavation should be carried out in two sections so that traffic flow is not completely disrupted. The usual safety precautions, as shown in Figure 3.5 must be observed.

Driving

Driving is a rather primitive method of forcing the service pipe under the road surface, and the degree of success depends on the type of fill used to form the road base. If rock is used as a road base, the chance of successfully driving the service pipe is remote as even the smallest piece of rock may cause the driving point to deflect from the desired line.

The equipment required consists of a driving point, driving rods and a striking head, assembled as in Figure 3.9. Rod guides are also used to retain the driving rod in the desired position and to correct alignment. After the main has been excavated, a trench is dug on the opposite side of the road at 90 degrees to the main, of sufficient length to accommodate the driving rod, and at a depth that will give the required cover to the service pipe.

The rod guides are then driven into the bed of the trench, aligned and levelled, and the driving rod assembled in the guides. To drive the rod under the road, the striking head is struck with a sledgehammer. After one length of rod has been driven up to the second rod guide, the striking head is removed and another length of rod attached to the previous length. This procedure is repeated until the driving point breaks through at the main excavation (see Figure 3.10).

When driving is complete, the rod is removed using a puller of the type shown in Figure 3.11. To prevent the hole from collapsing during or after withdrawal of the rod, the service pipe should be inserted from the water main side as the driving rod is withdrawn.

Jetting

Jetting can only be employed where the filling beneath the road consists of sandy or loamy soils. The equipment required is similar to that used for driving except that the jetting point is fitted with holes to allow the water to flow under pressure and erode away the soil, and the service pipe itself may be used to convey the water to the jetting point.

Water supply authorities and local councils are not in favour of this method as it wastes large quantities of water, and in loose sand and soils it is likely to create excessive

FIG 3.9 Driving equipment used for boring under narrow sections, such as culverts, footpaths, existing drains

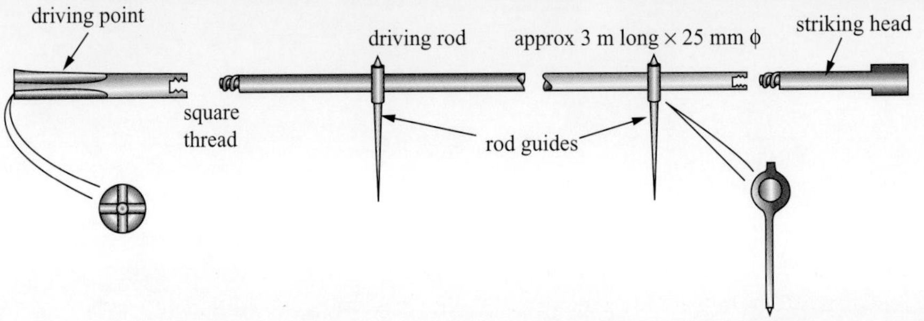

driving point

driving rod   approx 3 m long × 25 mm φ   striking head

square thread

rod guides

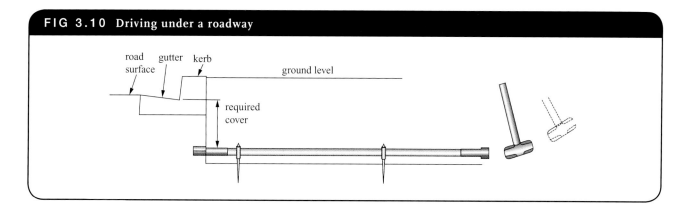

FIG 3.10 Driving under a roadway

FIG 3.11 Pipe puller—as the crowbar is moved backwards, the jaw grips the rod to remove it from the bored hole

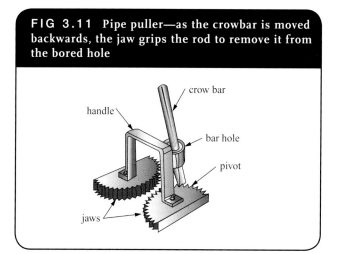

erosion which may cause the road surface to subside under heavy traffic or vibration. It is necessary to have the main drilled and tapped and the service control valve fitted prior to the laying of the service pipe.

Boring

The boring method of installing services beneath surfaced roads is by far the most commonly used, even though it has some distinct disadvantages in comparison to other methods. The main disadvantage is the inability of the operator to determine the difference between the various materials used for underground services. This often results in holes being bored through underground conduits conveying potentially dangerous substances such as electricity and gas. For this reason, extreme care should be exercised during the boring operation and the precise location of other underground services in the immediate vicinity must be known before boring commences.

Operation

After the water main has been excavated, a trench on the opposite side of the road is dug to the required depth to receive the boring cradle. The bed of the trench is levelled and the boring cradle positioned and anchored. The hydraulic pump motor is started, and the hydraulic motor rotates the boring rod and auger to penetrate for a distance equal to one length of the boring rod. Service pipes of

varying diameters may be installed using this method, by increasing the diameter of the auger.

Electrical safety as stated in AS/NZS 3500.1:2003

Safety precautions need to be observed when cutting into pipework or disconnecting water meters, fittings and devices on pipework. There have been fatalities and injuries that have been attributed to water services carrying an electrical current.

Where existing metallic service pipework is to be replaced in part or in its entirety by plastic pipe or other non-metallic fittings or couplings, the work should not commence until the earthing requirements have been checked by an electrical contractor and modified if necessary. Close attention to the requirements of AS/NZS 3500.1 and the local water authority is of the utmost importance.

MATERIALS FOR DOMESTIC WATER METER SERVICES

Domestic water services are generally laid using copper or metric polyethylene pipes and fittings in diameters from 20 mm to 40 mm. Copper tubes must be in accordance with national standards, as must polyethylene pipes. Fittings used on these materials must also comply with the relevant standards.

Copper and copper alloy pipes and fittings are not normally susceptible to external corrosion. However, external corrosion may occur when these materials are laid through filled ground containing ashes, salt (sodium chloride) or magnesite (magnesium oxychloride). Where the laying of pipes in these areas is unavoidable, pipes should be wrapped or coated externally with a suitable material capable of resisting the attack of these substances.

In unsewered areas, the contents of sullage pits and effluent absorption trenches may also accelerate external corrosion. Pipes in these areas should also be treated against external attack.

Metric polyethylene pipe

Polyethylene pipe is made from flexible PE80B medium density polyethylene material. It is highly cost effective and flexible for easier handling. With high-impact capabilities and UV resistance, it is manufactured to AS/NZS 4129.

FIG 3.12 Joint water supply system

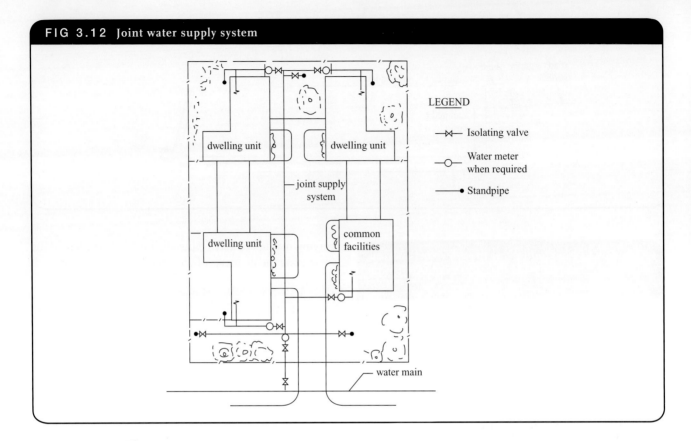

LEGEND

⊳⊲— Isolating valve

—○— Water meter when required

——• Standpipe

dwelling unit

dwelling unit

joint supply system

dwelling unit

common facilities

water main

Water service controls

All water supply services must be fitted with a service control valve, which is installed in the main at the point of connection with the service. The valve must be of an approved type acceptable to the water supply authority and suitable for installation below ground. Valves designed for this purpose are usually of a loose valve screw-down pattern, with the valve head securely attached to the body to prevent accidental removal of the head while the valve is under pressure.

Joint water supply controls

A 'joint water supply' is a term used for any privately owned water supply pipe that serves two or more dwelling units. Each dwelling unit and common facility used by the occupants of all units must be supplied by a separate branch from the joint supply pipe, and each must have its own isolating valve, positioned so that they can be isolated without affecting the water supply to other units or common facilities (Figure 3.12).

When this type of installation is required, an isolating valve of the loose screw-down pattern is installed in an accessible position as close to the property alignment as possible and fitted in such a way that all branch connections from the joint supply can be isolated from the water main.

WATER METERS

As a general rule, all domestic water supply installations must be fitted with a meter to record the quantity of water passing through to the consumer. They are usually installed within the property alignment in an accessible but protected position.

Selection of water meters

The selection of the size and type of water meter will be dependent on the required flow rates nominated by the applicant and the intended use of the development. All water meters used by a water corporation for billing purposes are to be of an approved type supplied by the water corporation. There are two basic types of meter in common use: volumetric and turbine.

Volumetric

The volumetric meter is a positive displacement type meter, also referred to as a piston meter. The meter incorporates a small bucket or piston with a known volume. The water flows through one orifice and fills the bucket, which then rotates and lets the water flow out through a second orifice. The number of buckets of water is converted to a total volume by the counter gearing.

VOLUMETRIC METER

Advantages:

- robust design brass body
- excellent low flow accuracy
- suitable for installation in any orientation
- suitable for water temperatures to 30 °C
- evacuated and sealed counter ensures clear reading.

Applications

Applications for volumetric metres include households and smaller commercial installations, leakage detection, and measuring flow into break tanks.

NOTE: This type of meter is generally susceptible to wear in poor water quality areas. However, the new style of volumetric meter offers superior wear resistance. High-grade engineering plastic is incorporated with the internal components, and processed using a unique moulding technique.

Turbine

With a turbine water meter, the water enters the meter body and is directed to a chamber around the measuring insert. The water then flows to the turbine via multiple passages spaced at intervals around the circumference of the insert. These passages form the 'multiple' jets of water that act to rotate the turbine. This rotation is transferred to the counter dials via the counter gearing.

TURBINE

Advantages:

- robust fully brass body and headring
- wet dial counter ensures clear reading
- direct drive from turbine to counter
- uniform accuracy over the whole of the meter life
- multi-jet configuration ensures even bearing wear
- suitable for water temperatures up to 50 °C
- excellent resistance to the impurities in water
- silent operation.

Applications

Turbine meters are used for general purpose metering (for small households and medium-sized commercial properties), and industrial installations where robust construction is required.

NOTE: Multi-jet turbine meters are typical of the meters used throughout Europe and many parts of the world. The multi-jet design is renowned for its long life and its resistance to the effects of poor water quality. To achieve their stated accuracy, multi-jet meters must be installed horizontally.

Water meter reading

Turbine meters differ from volumetric meters not only in the measuring principle they utilise but also in the way they display their measured volume. Volumetric meters typically have a single row of digits to display the reading. The first four digits are in black and represent whole cubic metres. The final four digits are in red and represent the fractions of a cubic metre.

Turbine meters use a combination of in-line digits and clock dials to show the volume of water that has been measured. The inline digits are in black and show the whole cubic metres. The clock dials have read points that indicate the fractions of a cubic metre.

Installation

Water meters must be installed level, approximately 75 mm above ground (Figure 3.13). The position selected for the meter should be within the property alignment and accessible at all times for reading and maintenance.

Where meters are installed in driveways and may be subjected to vehicular damage, they may be recessed into walls or surrounded by a suitable guard to prevent damage. Under normal circumstances meters should not be installed below ground. However, in areas where the temperature may drop to below freezing, the meters may be installed below ground, subject to special approval, in a properly constructed and drained box (Figure 3.14).

Meter controls

Every meter is required to have a control valve fitted at the inlet. This control is of the loose valve screw-down pattern and incorporates a 'T' head or similar suitable for hand operation. Some authorities allow for a right-handed, lockable ball valve to be installed, providing the minimum backflow requirements are met.

On installations that involve excessive lengths of outlet pipework from the meter, or on multi-storey installations, an additional control valve may be fitted at the outlet of the

FIG 3.13 Water meter installed above ground

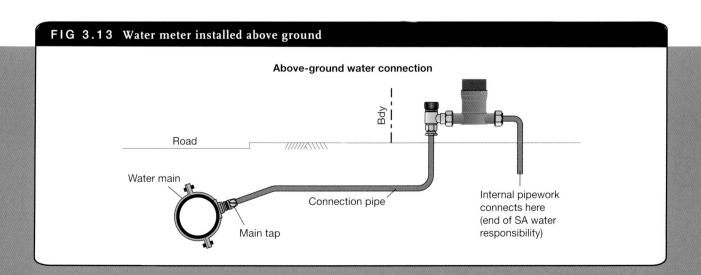

Above-ground water connection

Road

Bdy

Water main

Connection pipe

Main tap

Internal pipework connects here (end of SA water responsibility)

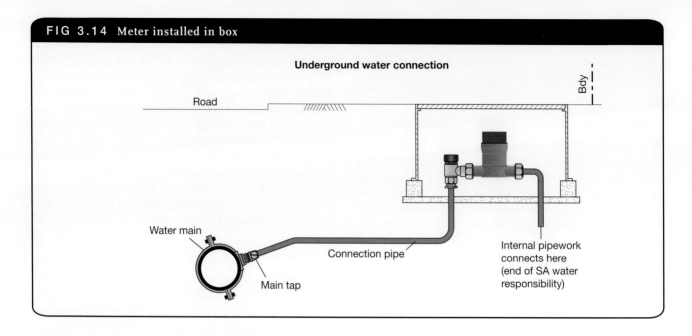

FIG 3.14 Meter installed in box

Underground water connection

Bdy

Road

Water main

Connection pipe

Main tap

Internal pipework
connects here
(end of SA water
responsibility)

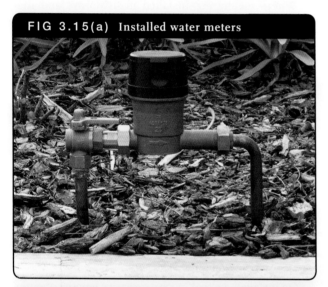

FIG 3.15(a) Installed water meters

FIG 3.15(b) Installed water meters

meter. If the meter is disconnected, this valve prevents the backflow of water contained in the pipework which could create a hazardous situation.

Backflow prevention

Some states and territories require all new connections and redevelopments to have an appropriate backflow prevention device fitted at the outlet of the main water meter (containment protection) in accordance with plumbing regulations incorporating the Plumbing Code of Australia.

If the risk category of a non-residential development is unknown at time of application, the water corporation may require the installation of a high-hazard backflow prevention device. For single residential properties, a low-hazard dual check valve may be required to be installed at the outlet of the water meter (Figures 3.15 (a and b)).

NOTE: Where the installation of an appropriate zone or individual hazard backflow prevention device is necessary in accordance with the provisions of AS/NZS 3500.1:2003, the relevant water corporation may require, as a minimum, the same level of protection installed as a containment backflow prevention device at the outlet of the property main water meter.

Where above-ground rainwater tanks are installed to provide toilet flushing, and it is intended to interconnect the reticulated drinking water supply system from the relevant water authority, an appropriate containment backflow prevention device will be required at the outlet of the main water meter to the property. In such cases, as a minimum, the device is to be a watermark approved, dual check valve.

SUSTAINABLE PLUMBING: RECYCLED WATER

Recycled water is water taken from any waste stream and treated to a high standard so it can be used for a new activity. Recycled water can refer to fully treated effluent from sewage treatment plants.

Recycled water is a secure alternative water source that, when treated as required, is fit for a range of purposes, such as:

- agricultural irrigation
- industrial processing such as for cooling
- municipal uses such as watering parks and gardens
- domestic uses such as toilet flushing, car washing and garden watering.

As an example, an extensive system of pipes and aqueducts distributes water from Melbourne's water storage reservoirs to the retail water companies and their customers.

National guidelines, such as Australian guidelines for water recycling, require the water to be 'fit for the intended purpose'. Environmental factors, such as salinity and nutrient levels, also need to be considered to ensure recycled water is suitable for the intended use.

Upgrades to sewage treatment plants have improved the quality of recycled water and made it suitable for a wider range of uses. Water companies, in conjunction with water industry partners and the government, are continuing to develop programs to support recycled water to be used by farmers, industry, local councils and households.

NOTE: It is important to match the quality of recycled water to its intended use.

FOR STUDENT RESEARCH

Investigate and report on approved materials (not those covered in this text) for water services in your area. Include in your report any specific requirements for these materials.

Australian Standards

AS/NZS 3500.1 Plumbing and drainage—Water services

AS 3565 Meters for water supply

AS/NZS 2845 Water supply–Backflow prevention devices

AS/NZS 1477 PVC pipes and fittings for pressure applications

AS/NZS 2032 Installation of PVC pipe systems

AS/NZS 4441 PVC-O pipes for pressure applications

AS/NZS 4765 Modified PVC (PVC-M) pipes for pressure applications

4 Materials, jointing and fixing for hot and cold water supply

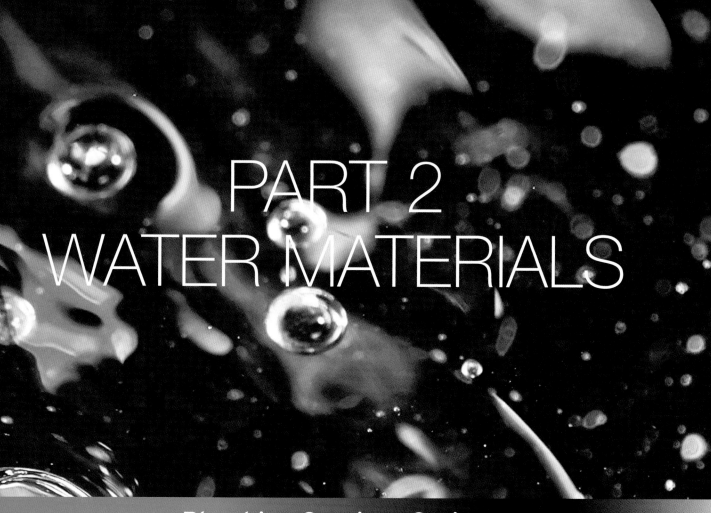

PART 2
WATER MATERIALS

Plumbing Services Series

CHAPTER FOUR

Materials, jointing and fixing for hot and cold water supply

LEARNING OBJECTIVES

In this chapter you will learn about:

INTRODUCTION

In most states water conditions play a significant part in choosing the type of material to be used. Installations should also be aesthetically pleasing, so that all pipework installed in buildings should be concealed where possible. The most common materials used in hot and cold water services are:

1 copper

2 galvanised mild steel

3 stainless steel

4 PVC-U

5 polyethylene.

COPPER

Copper tubes are seamless. In the first stage of production, copper is made into a rough tube shell, either by a rotary piercing process or by extrusion. It is then subjected to a series of cold drawing operations, during which the copper hardens. It has to be softened by annealing at various stages between drawing operations. The tube hardness can be altered to suit the particular purpose for which it will be required—hard, half hard and soft tube can be produced by adjustment of the last or final drawing and annealing process to which the tube has to be subjected.

Copper tubes in hot and cold water services are usually in a half hard state or temper. They are supplied in lengths averaging 6 m and have sufficient rigidity and straightness to prevent any damage in initial packaging and transit to the job. Tubes in a soft or annealed state are supplied in longer lengths and in coils for convenience for handling and transport. Copper tubes are measured by their external diameters, the wall thickness varying depending on the purpose for which the tube is to be used.

Jointing copper tube

The recommended jointing procedures for light gauge copper tubes in hot and cold water lines include capillary fittings, compression fittings and crimped fittings, and the use of jointing tools to form sockets, tees and bends in the tube, which are jointed by silver soldering or brazing.

Capillary fittings

These fittings are made from copper and copper alloys (Figure 4.1). Copper fittings are usually extruded cold wrought from copper tube; wrought copper fittings are neat, strong and free from porosity. Copper alloy fittings are made by casting or hot pressing. These fittings are

FIG 4.1 Capillary fittings; some copper-to-brass soft-solder joints

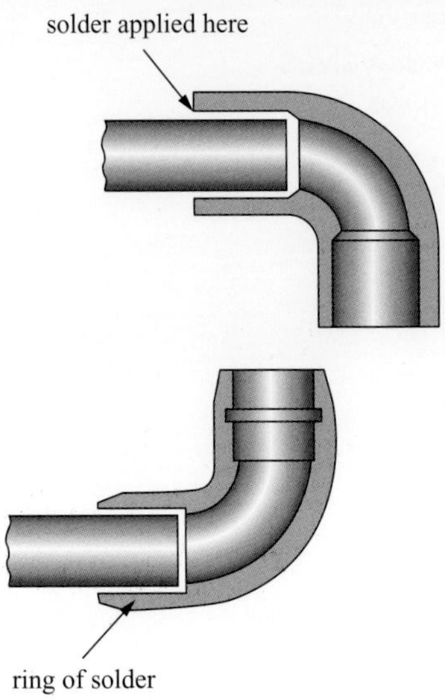

solder applied here

ring of solder

said to be much more free from possible porosity, but in some areas of different soils, they may suffer from dezincification (see Glossary), and in these areas a cast gunmetal or wrought copper fitting is used.

Capillary fittings have sockets made to close tolerances so that a controlled gap exists between the tube end (or spigot) and the socket, into which molten soft solder is drawn by capillary action. (Note that this type of joint is not permitted in South Australia.) The solder may be incorporated in the fitting by means of a ring of solder. The steps required to make a successful capillary joint are described in the procedure box.

PROCEDURE FOR MAKING A CAPILLARY JOINT

1. Cut the tube square, with either a fine-toothed hacksaw or tube cutter.

2. Remove the burr created by the cutting tool.

3. Clean the tube end and fitting socket with steel wool, not an abrasive or emery paper as this could affect the neat fit, resulting in a poor joint.

4. Apply flux to the tube end and fitting socket.

5. Insert the tube fully up to the stop in the fitting socket and wipe off the excess flux.

6. Apply heat with a propane (LPG) torch, air acetylene or oxyacetylene torch. The flame should be played on the fitting and kept moving to heat the whole joint area. This will avoid local overheating. Overheating can burn the flux, destroying its effectiveness, and can cause cast fittings to crack.

7. With the integral ring of solder type of capillary fitting, heating is continued until a complete ring of solder appears around the mouth of the fitting. The heating is then stopped and the joint is allowed to cool without disturbance and then cleaned.

Fixing or clipping of copper tubes

Copper tubes should be secured in such a way that they will not be damaged or deformed or prevented from expanding or contracting. The distances between fixing points for horizontally and vertically run tubes should not exceed those in Table 4.1. (See also AS/NZS 3500.1 Table 5.2 available from Standards Australia, or through your TAFE, school or college library for further pipe diameters.)

There are various types of clips and brackets to meet specific requirements. Manufacturers' catalogues illustrate a selection, and this information will help the plumber to decide on the best type to use for a particular job.

Copper hot and cold services embedded in concrete (structural slabs)

Copper has excellent resistance to corrosion and is not attacked by the normal types of cement, plaster or concrete.

TABLE 4.1 Maximum spacing of brackets and clips, in metres

Nominal pipe size DN	Copper and stainless steel pipes	Galvanised steel pipes
10	1.5	–
15	1.5	2.0
18	1.5	–
20	1.5	2.0
25	2.0	2.0
40	2.5	2.5
50	3.0	3.0

However, acid plasters, coke breeze or furnace slag and acid cements will attack the copper, causing porosity and need to be protected accordingly.

Compression joints

Compression joints used on copper tubes may be divided in two groups: non-manipulative and manipulative.

Non-manipulative joint

This type of joint is made by tightening the coupling nut against a grummet or sleeve to apply pressure to the exterior of the tube (Figure 4.2). It should only be used on half-hard tubing; if used on softened tube it could cause severe indentations on the tube, which results in a joint that appears sound but may leak later.

To manufacture the joint, the tube is cut square and the nut, followed by a small ring of special design, is slipped on to the tube. The nut is then tightened, compressing the grummet or ring, and forcing it into the circumference of the tube, causing a slight ripple or corrugation.

Manipulative joint

To make a manipulative joint, the tube is cut square and deburred, then flared, cupped or belled with special forming tools (Figure 4.3(a)). It is compressed by means of a coupling nut, against a shaped end of the corrugated section or internal cone (Figure 4.3(b)).

A different type of manipulative compression joint is one in which a special tool commonly called a 'crox'

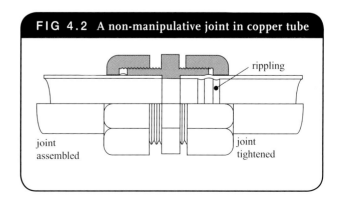

FIG 4.2 A non-manipulative joint in copper tube

rippling

joint assembled

joint tightened

FIG 4.3(a) Flared manipulative joint

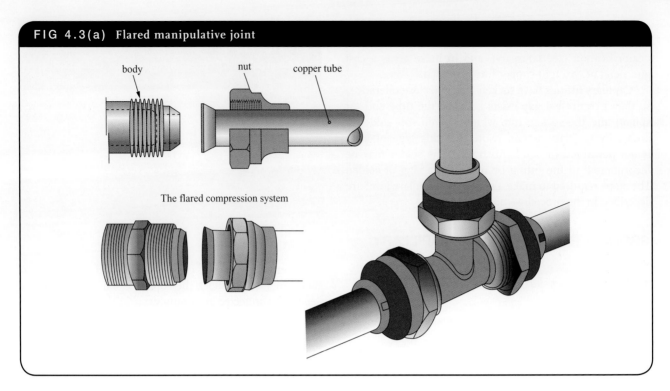

body   nut   copper tube

The flared compression system

FIG 4.3(b) Beaded manipulative joint

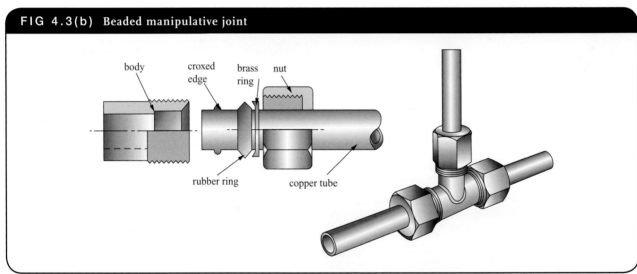

body   croxed edge   brass ring   nut

rubber ring   copper tube

tool is used to roll a bead on the tube about 12 mm from the end. When the coupling nut is tightened, the bead is compressed against the mouth of the fitting and makes a successful joint. Sometimes it is advisable to use a grummet of hemp or teflon tape to ensure the joint will not leak under pressure.

Silver-alloy brazing of copper tube

Silver brazing is the easiest and most satisfactory method of jointing copper to copper and copper to brass.

Capillarity plays an important part in this method of jointing. When the metal surfaces are clean and fit closely together, the brazing alloy flows into the joint by capillary action. A joint so made gives maximum strength and a small quantity of silver brazing alloy is used, whether a fitting is used (Figure 4.4), or whether a spigot and socket is formed by hand tools from the pipe ends (Figure 4.5). Silver brazing rods consist principally of copper, zinc and silver and are the best for this method of jointing. Silver brazing alloys with as low as 1.8 per cent silver are approved for use in hot and cold water services in most states.

Care should be taken not to overheat the alloy as some of the properties are easily destroyed. On the other hand, if it is underheated, the alloy will not penetrate the molecular structure of the copper tube and fitting surfaces. This will result in a rough, weak and unreliable joint.

Joint preparation of copper tubes using silver brazing alloys should be carried out in the same manner as that

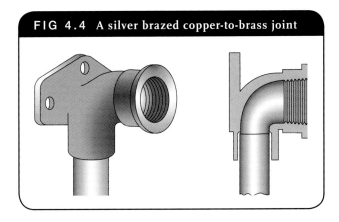

FIG 4.4 A silver brazed copper-to-brass joint

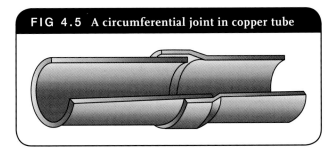

FIG 4.5 A circumferential joint in copper tube

adopted for capillary joints. The joint is heated evenly and the brazing alloy is applied to the edge of the joint. If the joint is not heated sufficiently, the brazing alloy does not flow readily when it is applied to the joint. Flux is not required on copper-to-copper joints.

GALVANISED MILD STEEL TUBE

Ordinary screwed pipe is generally of mild steel with a longitudinal seam, coated inside and outside with zinc after manufacture. This tube is also known as wrought iron pipe. It is used extensively throughout Australia in cold water, fire and gas services. This type of pipe is measured by its inside diameter (ID).

Jointing galvanised steel tube

The most common method of joining galvanised steel pipes is by the use of threaded joints, which may either be parallel female and taper male, or parallel male and female threads.

Parallel female and taper male threads

This is the most common type of joint used. The taper on the male thread ensures that complete contact of the metal surfaces occurs at some point and further screwing beyond this point will produce a watertight joint.

Jointing compound, teflon tape or hemp packing is used to fill the small space between the female and male threads. Hemp is still used extensively in the fire protection industry. It is usually purchased in rope form and is a reliable jointing material. The hemp is

teased out to a length equivalent to about three times the circumference of the tube. One end is applied to the last two or three threads at the end of the tube. Then the hemp is wound tightly on the male thread, making sure that it is wound in the same direction as the female thread travels as it is turned on the male thread. Excess hemp will be pushed off the male thread as the joint is tightened. After the joint is made, any excess hemp on the exposed threads should be removed. Teflon tape is applied using the same method.

Parallel female and parallel male threads

As the name suggests, these threads are formed within a parallel sided fitting in contrast to the usual taper fitting. A typical parallel thread fitting is the longscrew or connector that is used to join two fixed pipelines. Sealing is by way of a hemp grummet and a backnut.

Galvanised pipe fittings

Fittings available for use in mild steel hot and cold water services are galvanised and are generally of malleable cast iron (Figure 4.6).

STAINLESS STEEL TUBE

Stainless steel tube is one of the most corrosion-resistant materials available to plumbers for use in hot and cold water services. This is due to the resistance of the hard, adherent and transparent oxide film that covers the surface of the tube.

Stainless steel tube is stronger than copper and steel and weighs less. The rigidity of the tube is an advantage when fixing, as is the expansion rate of stainless steel tube, which is less than that of copper and results in less movement in fixing and less strain on joints. The tube is fully compatible with copper or copper alloy fittings; no galvanic or electrolytic action occurs. The stainless steel tube can be embedded in brickwork or plaster without fear of corrosive attack. Similar precautions to those used with copper tube should be taken when installing in concrete or a structural slab. It is recommended that stainless steel tubing should not be used underground except where it is protected against external corrosion by lagging or conduit.

Jointing stainless steel tube

Stainless steel tube can be joined in the same way as copper tube, with capillary fittings, silver brazing and compression fittings. Suitable capillary fittings, such as end feed fittings and integral solder rings, can be used. The flux required is a phosphoric acid based paste flux. The area to be soldered should be clean and free from oil and grease, the tube end and capillary fitting should be cleaned with emery paper, and the paste flux applied to both. Heat is applied using a propane torch (LPG), air acetylene or oxyacetylene torch. The heat should be applied evenly around the fitting with sweeping movements, avoiding heating the tube because of the low heat conductivity of stainless steel.

FIG 4.6 Wrought iron and malleable cast iron pipe fittings

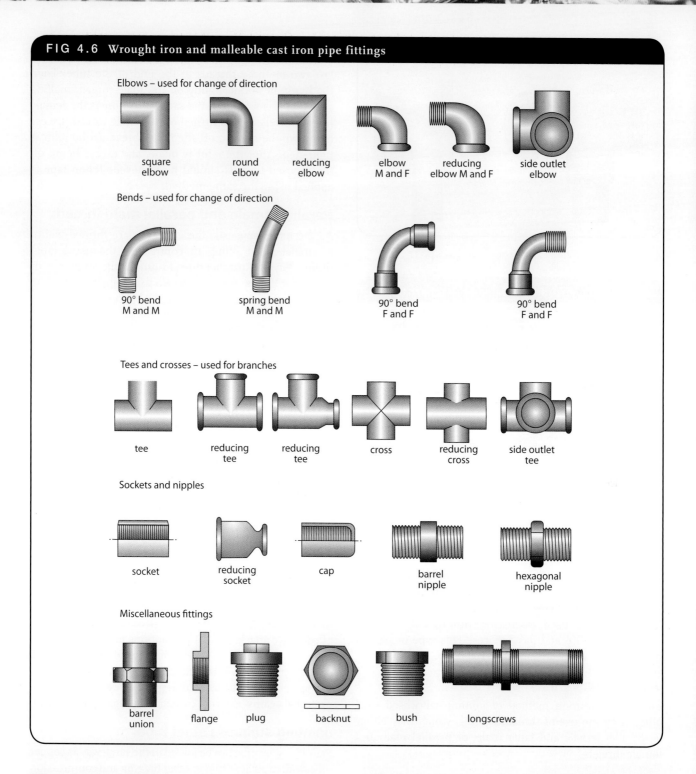

Elbows – used for change of direction

square elbow   round elbow   reducing elbow   elbow M and F   reducing elbow M and F   side outlet elbow

Bends – used for change of direction

90° bend M and M   spring bend M and M   90° bend F and F   90° bend F and F

Tees and crosses – used for branches

tee   reducing tee   reducing tee   cross   reducing cross   side outlet tee

Sockets and nipples

socket   reducing socket   cap   barrel nipple   hexagonal nipple

Miscellaneous fittings

barrel union   flange   plug   backnut   bush   longscrews

Silver brazing joints

Brazed joints are prepared in the same way as soldered joints. A general purpose silver brazing flux should be applied immediately to both surfaces after cleaning. Heat should be applied evenly around the fitting, removing the flame occasionally to allow the heat to spread evenly throughout. The silver rod is applied to the mouth of the fitting when the rod's melting point is reached.

A socket joint can be formed with a mechanical expander. These joints must be silver brazed. If an oxide has formed because of excessive heat being applied, the joint cannot be completed until the oxide is removed.

Compression joints

Compression joints can be used with stainless steel tube. However, it is essential that the tube to be flared is cut with a hacksaw, not a disc cutter.

Fixing or clipping

Pipe clips or brackets may be of various materials. If made of dissimilar metals, they should be covered with a suitable coating to prevent contact with the stainless steel tube. When it is necessary to insulate hot and cold water lines, normal insulation materials are quite satisfactory. Because of the presence of chlorides (which cause corrosion) in styrene foam, care must be taken to keep the insulation material dry during and after installation.

UNPLASTICISED POLYVINYL CHLORIDE (PVC-U)

Unplasticised polyvinyl chloride is a tough rigid material available in a large range of colours, and is one of the most widely used of the plastic pipe materials. PVC-U consists basically of ethylene and hydrogen chloride, which are derivatives of petroleum and salt and water. It is a stable, strong and relatively inexpensive material, and has excellent ultraviolet light resistance when a chemical compound such as titanium dioxide, which absorbs ultraviolet rays from the sun, is added. Without the addition of the chemical compound to protect the PVC-U surface, crazing occurs and the pipe becomes brittle, leading to a breakdown of the pipe and leaks.

PVC-U was developed during World War II to provide a substitute for corrosion-resistant alloys such as stainless steel, which were in short supply. It is a good heat insulator and has a high coefficient of expansion, approximately six times that of steel. It is not recommended for situations where temperatures above 60°C or below 0°C are experienced; it may however be handled at quite low temperatures approaching 0°C. PVC-U pipe is made in an extruded form and, because of its rigidity, tubes are not supplied in coils but in 6 m lengths.

Jointing PVC-U

Joints in PVC-U are fairly simple to make and, providing the proper procedure is followed, little or no trouble will be experienced when this material is used in water service pipelines. The most common methods used are welding and solvent cementing.

Solvent cementing

A large range of socket fittings is available in Australia for PVC-U installations. They must be of the correct quality and able to withstand in excess of the maximum working water pressure in the system. The piping and fittings must be clean and dry, and the manufacturer's instructions followed for all jointing procedures (Figures 4.7 and 4.8).

SOLVENT CEMENT JOINTS

Advantages:

- no specialised equipment is necessary.
- joint becomes stronger than the parent metal.
- joint is easily and quickly made.
- valves, metal fittings and taps may be screwed into special PVC-U sockets.
- sockets may be formed on the ends of straight lengths of pipe and the joint solvent cemented.

FIG 4.7 Preparation of solvent cemented joint: (a) roughen with emery paper and clean, (b) degrease with methylene chloride, (c) brush on cement, (d) push tube into fitting

(a)   (b)

(c)   (d)

FIG 4.8 Completed solvent cement joint

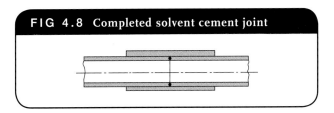

Fixing of PVC-U pipelines

When fixing PVC-U pipeline, reference should be made to the appropriate national Standard AS/NZS3500 for such specifications as fixing spacing, expansion joints, sliding supports, and building into floors and walls. Refer to Table 4.2. View AS/NZS 3500.1 Table 5.2 (available from Standards Australia, or through your TAFE, school or college library) for further pipe diameters.

TABLE 4.2 Maximum spacing of brackets and clips, in metres

Nominal pipe size DN	PVC-U, polyethylene, cross-linked polyethylene, polypropylene	
	Horizontal or graded pipes	Vertical pipes
10	0.50	1.00
15	0.60	1.20
16	0.60	1.20
18	0.60	1.20
20	0.70	1.40
22	0.70	1.40
25	0.75	1.50
40	0.90	1.80
50	1.05	2.10

Electrical earthing

Being of a non-conductive material, PVC-U pipe must not be used as a means of earthing electrical installations or of dissipating static charges. Electricity supply authorities must be advised by the installer where any existing metallic water supply pipelines are being replaced by PVC-U pipe. This action could prevent an electrical hazard for a person inadvertently using electrical equipment that has not been earthed in accordance with the regulations.

POLYETHYLENE (PE-X)

Cross-linked high-density polyethylene (PE-X)

Cross-linked high-density polyethylene (PE-X) is simple and quick to install and has proven to be very reliable and saves time, labour and material costs. The strength and flexibility of PE-X pipes used together with a range of dedicated dezincification-resistant (DZR) brass compression sleeve fittings makes this a superior system that can be used in residential and commercial applications, for hot and cold water services, rainwater services from tanks into the buildings and even to distribute recycled or reclaimed water from the mains into the buildings, using the following colour codes:

- green—rainwater
- lilac—recycled water
- red—hot water
- yellow—gas is specifically designed and tested for LPG and NG application in Australia and New Zealand.

The cross-linking process causes the individual polyethylene molecules within the piping to bond in a three-dimensional network, creating a tough and durable pipe with:

- long-term stability
- high temperature resistance
- high pressure resistance
- superior flexibility and strength.

Reduced water hammer

Water hammer commonly occurs when there is sudden stop of water flow in the plumbing system, for example when closing a tap or shutting off a water valve. Water hammer not only creates unpleasant noise, it can cause extensive property damage as the pressure surge can sometimes exceed 60 bar, leading to rupture of compromised piping and fittings.

Due to the elastic behaviour of the cross-linked polyethylene, water hammer effects can be reduced by up to 75 per cent compared with similar metallic systems.

Pressure testing

Pressure testing is conducted before concealment. The pressure hydraulic water test must be conducted in accordance with AS/NZS3500, which is pressure tested to 1500 Kpa for a minimum of 30 minutes.

Jointing cross-linked high-density polyethylene (PE-X)

Because of the composition of polyethylene, no effective commercial solvent has yet been developed that will dissolve or fuse it. Solvent joints as used with PVC-U are therefore not possible. Jointing is carried out by means of the use of brass or polymer compression fittings (Figure 4.9).

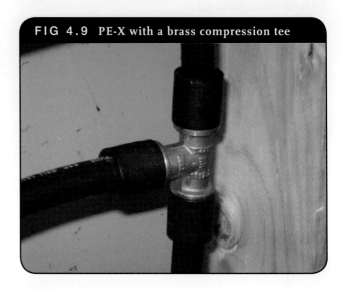

FIG 4.9 PE-X with a brass compression tee

Fixing

The coefficient of expansion of polyethylene is high; the material should therefore be free to expand and contract wherever it is fitted. The fixing of the pipe is shown in Table 4.2.

POLYETHYLENE (PE)

Polyethylene (PE) pipe is made of a colourless thermoplastic that remains flexible over a wide temperature range. It is a synthetic chemical material manufactured by polymerisation of fractions of low molecular weight petroleum or natural gas. Carbon black is added during the production to increase resistance to the sun's ultraviolet light.

There are two types of PE. Black polyethylene should always be used in water plumbing; white polyethylene has no stabilising agent to prevent deterioration of the pipe. White polyethylene is used mainly in underground sprinkler systems and for rural and agricultural purposes. Long lengths may be laid underground by the use of a mole plough and tractor, or in trenches.

The main characteristics of polyethylene are:

- The material has a high rate of expansion and contraction.
- There is little rigidity in long length and it is produced in coils of specified lengths.
- The material can be bent to any shape after heating by hot water.
- The pipe has a smooth mirror finish bore, which is a characteristic of most plastic pipes.

FIG 4.10 A polypropylene connection to polyethylene pipe

FIG 4.11 Remember to use the correct cutting technique to ensure the pipe is free from burrs

Polypropylene fittings

Polypropylene fittings include a range of mechanical joiners, elbow and tees ready-assembled to easily join blue line pressure poly pipe for cold water applications to meet Australian standards (Figures 4.10 and 4.11)

The metric size fittings incorporate advanced thermoplastic materials to provide impact, UV, chemical and corrosion resistance. They are pressure rated to 1600 kPa to meet the needs of high-pressure systems. The fittings are also designed for fast and simple connection to PE, PE-X, PVC-U and copper pipe.

Metric compression fittings for recycled water

The metric compression fittings ranges for dedicated, recycled water are made with bright purple nuts to clearly identify a recycled water pipe connection. This greatly reduces the risk of a cross connection between recycled and potable water systems in dual reticulation applications.

Electrofusion

Electrofusion technology is an alternative method for joining polyethylene pipe. Manufacturers offer a specialised range of fusion fittings that ensure each connection is superior, reliable, simple and safe. Fittings for the electrofusion method are equipped with exposed heating coils. These ensure symmetrical heat transfer between the fitting and pipe during fusion.

The butt welding process requires training to gain the expertise required to obtain a guaranteed water tight joint.

Refer to the *Basic Skills* textbook, Chapter 5 for further explanation of butt and electrofusion welding.

FOR STUDENT RESEARCH

Investigate and document where recycled water distribution systems are being installed in your local area.

Australian Standards

AS/NZS 4129:2000 Fittings for PE pipes for pressure applications

AS/NZS 3500.1:2003 Plumbing and drainage—Water services

AS/NZS 3500.4:2003 Plumbing and drainage—Heated water services

AS/NZS 2642.2 Polybutylene pipe systems—Pipe

AS/NZS 2642.3 Polybutylene pipe systems—Mechanical jointing fittings

ON-SITE STORIES 4.1

CHECK, RE-CHECK ... AND CHECK AGAIN!

Martin Kupferle, 2nd year plumbing apprentice, Victoria

It all started with a Green Plumber course at the Plumbing Industry Climate Action Centre in Brunswick, Victoria. I attended a week's course on sustainability in plumbing. The centre is a wonderland of new technology: solar heating, tri-generation motors, water recycling, water conservation—the list goes on! All up, it's a 'wow' place. Having a better understanding of sustainable plumbing and the benefits to our customers, and armed with my new Green Plumbers qualification, I went back to work.

A month later I had my chance to show what I had learned. One of our customers was impressed by my renewable energy knowledge and gave us a job; twin hot water return lines with return pumps, rainwater tanks and a solar hot water service.

We arrived and did the installation. The solar hot water service was the icing on the cake! We took great care to follow the manufacturer's instructions to the letter, checked the unit we had, and off I went. We completed the commissioning checklist and all looked good. We started her up and everything was fine; the water reached temperature, there were no leaks. It was a piece of art! The boss gave me a pat on the back, the customer was rapt, all up great job!

A few weeks later we received a call from the customer, and after expecting it to be good news, my jaw dropped when they complained about how much gas they were using and how disappointed they were with it! So back we went, tail between the legs, and inspected everything. Not a water drip or problem in sight. I had done the best job possible. Then I had a thought … I referred to the manufacturer's instructions and checked and checked again. I checked the paperwork against the unit, which was all correct, but still it didn't work, much to my frustration! Then it hit me. The solar hot water system we received wasn't what we had ordered. Although it looked the same, it differed in the way the panels on the roof were connected! Half an hour later, I had swapped the feed line from the top to the bottom and bingo! A little red faced, but relieved, I left.

Six weeks later I bumped into our client at the supermarket. They raved about how little gas they were using and how much money they were saving! So my first sustainable plumbing job was a success. Hopefully it will lead to more sustainable work in the future.

And the moral of my story? Check and re-check. And when all else fails … check again!

CHAPTER FIVE

Valves, taps and controls

LEARNING OBJECTIVES

In this chapter you will learn about:

5.1 valves and taps

5.2 float valves

5.3 check valves

5.4 pressure and control valves

5.5 vacuum break valves

5.6 solenoid valves

5.7 hot and cold mixing valves

5.8 tempering valves

5.9 thermostatic mixing valves.

INTRODUCTION

Water supply systems require a number of mechanical devices to control and regulate the flow of water through the pipework. In some instances these devices also act as safety devices should some fault occur in the system or associated equipment.

The range of valves, taps and controls used in water supply work is immense. Each has been designed for a specific purpose and therefore the materials from which they are manufactured also vary considerably. Only those which may be found on domestic installations are covered in this text.

THE DIFFERENCE BETWEEN A VALVE AND A TAP

Valves

A valve is a hand-operated device used to control the flow of water in a piping system. It is manufactured with inlet and outlet threads, flanges, capillary joints or other approved jointing methods that enable it to be installed in an 'in-line' position.

Valves may be manufactured from bronze, brass, gunmetal, steel, cast iron, glass or plastic or a combination of these. Materials with high corrosive resistance are usually used for water supply installations.

Taps

A tap is also a hand-operated device used to control the flow of water in a piping system. A tap may or may not be manufactured with both inlet and outlet threads. Unlike a valve it is designed to be installed on an outlet of a pipeline and the type of appliance water outlet dictates the design of the tap.

GROUPS OF VALVES AND TAPS

There are several groups of valves and taps in common use for water supply. The types are:

- loose valve screw-down
- straight through screw-down
- ground-face plug
- ceramic disc.

Loose valve screw-down type

These are the most commonly used controls on reticulated water supplies and there are many different shapes in both valves and taps that incorporate the same operating principle (Figures 5.1, 5.2 and 5.3).

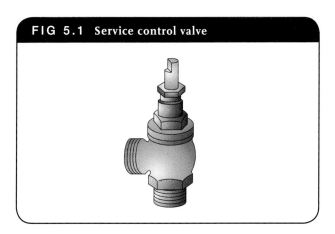

FIG 5.1 Service control valve

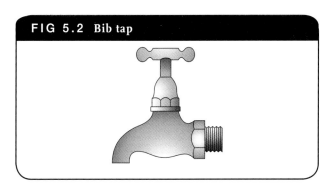

FIG 5.2 Bib tap

FIG 5.3 Loose valve screw-down valves and taps: (a) Meter control val ve, (b) F + F stop valve, (c) M + F stop valve, (d) hose tap, (e) pillar tap, (f) recess tap

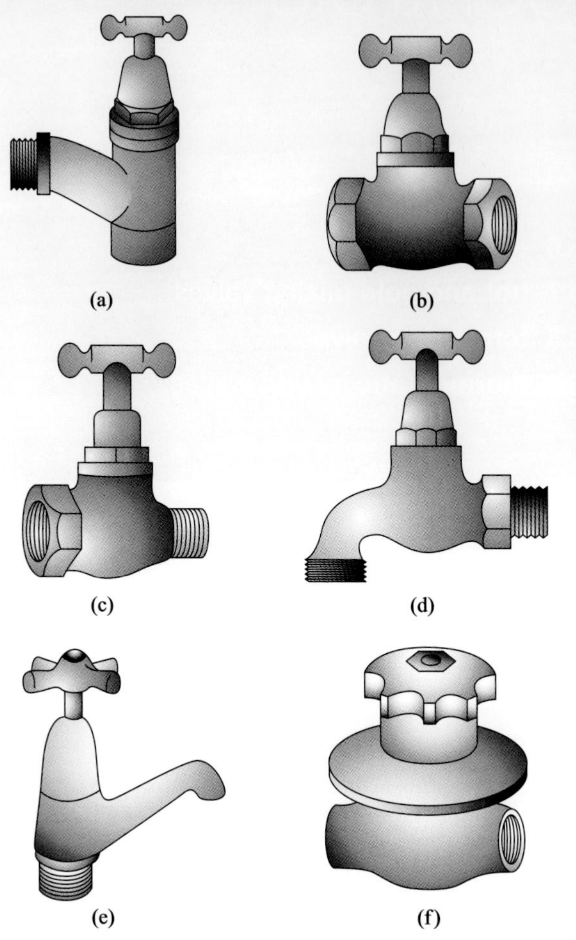

(a)

(b)

(c)

(d)

(e)

(f)

FIG 5.4 Operation of loose valve screw-down types: (a) open, (b) closed

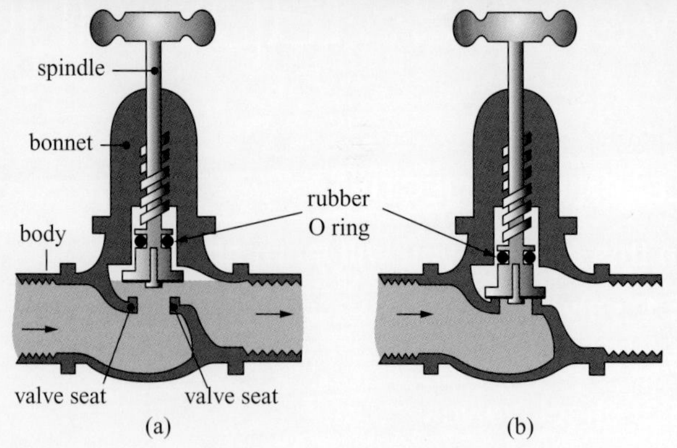

spindle

bonnet

body

rubber O ring

valve seat   valve seat

(a)

(b)

FIG 5.5 Installation of a loose valve: (a) correct, (b) incorrect

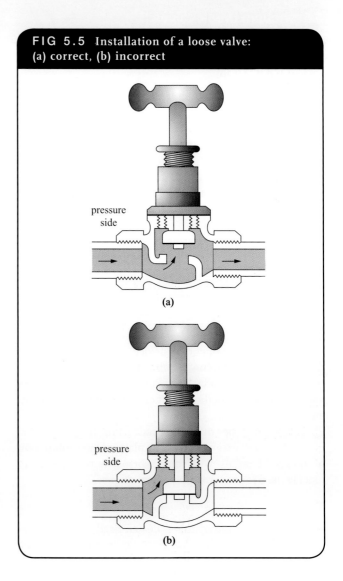

pressure side

(a)

pressure side

(b)

Operation

The screw-down spindle is designed so that the closing or opening of the valve is a gradual operation, which minimises noise and vibration caused by water hammer. A loose jumper valve is forced down on to the valve seat by screwing down the spindle. The pressure exerted on the valve stops the water passing through it (Figure 5.4).

When the spindle is raised by turning the handle in the anti-clockwise direction, pressure is released from the valve and the water pressure acting on the underside is then able to lift the valve and allow the water to pass through to the outlet.

Water is restricted to the body section by a rubber O-ring fitted on the bottom of the spindle. This O-ring forms an effective seal between the spindle and the bonnet of the valve and prevents water passing out of the tap via the bonnet.

Some valves and taps are not fitted with O-rings but have a stuffing box that retains a graphite-impregnated asbestos gland serving the same purpose.

All loose valve taps and valves have, as their name implies, a loose valve to shut off the water supply. This loose valve allows the water to flow through in one direction only. Therefore it is necessary to install this valve so that the incoming water lifts the valve from beneath. If the valve is installed the wrong way round, the opposite will occur and water will be unable to pass through the valve (Figure 5.5).

The loose jumper valve may be manufactured of brass and have a replaceable neoprene washer. Alternatively the jumper valve may be a plastic throwaway type with an integral washer (Figure 5.6).

Straight through screw-down types

Commonly referred to as gate valves, these are manufactured with a straight through passage for the flow of water. When fully opened, there is little or no restriction of the flow through the valve, as is the case with the loose valve screw-down type (Figure 5.7).

Operation

Unlike in loose valve types, the spindle of the gate valve is fixed and does not raise and lower as the valve is operated. The tapered gate has an internal thread that matches the external thread of the spindle, so when the hand wheel is operated, the gate is raised or lowered by the two threads. The water flow is controlled by forcing the wedge-shaped gate between the tapered seats of the valve.

As water pressure is not required to operate this type of control, it may be installed with the inlet on either side.

The use of the gate valve in domestic water supply is generally restricted to drain or sludge valves on storage tanks. However, it is used extensively on transmission and distribution pipework.

Ground-face plug valves

These are rarely if ever used on reticulated water supply. However, they are indispensable for gravity-fed rural water supply systems. They were first used hundreds of years ago and have remained unchanged. The only major variation over this period is the materials from which they are manufactured.

Ground-face plug valves consist of a tapered plug fitted to a matching tapered body (Figure 5.8). The valve relies on metal-to-metal contact to prevent leaks, although some modern types are fitted with neoprene O-rings.

FIG 5.7 Operation of gate valve: (a) valve showing hand wheel, (b) open, (c) closed

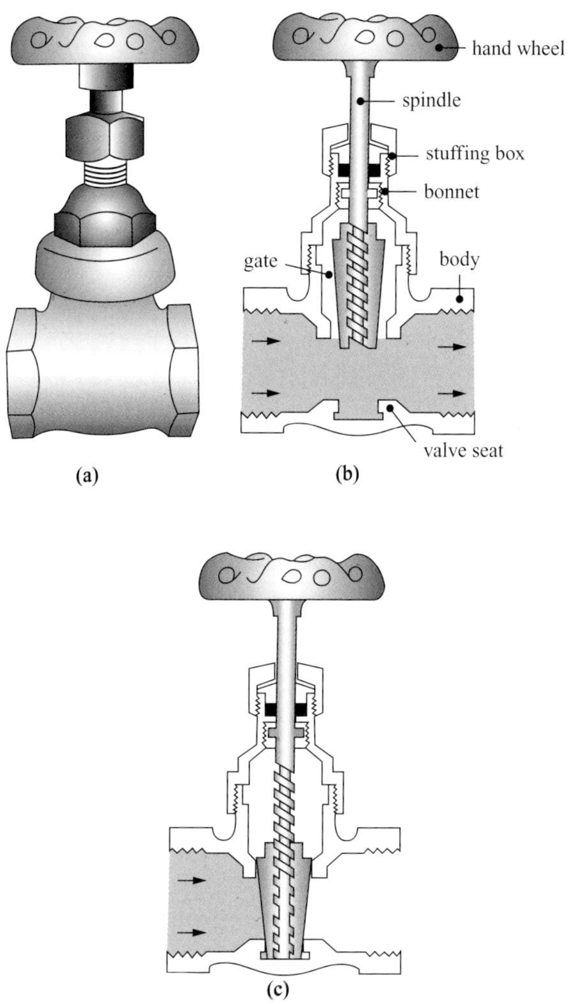

hand wheel
spindle
stuffing box
bonnet
gate
body
valve seat
(a) (b)

(c)

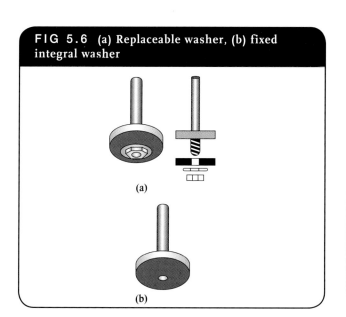

FIG 5.6 (a) Replaceable washer, (b) fixed integral washer

(a)

(b)

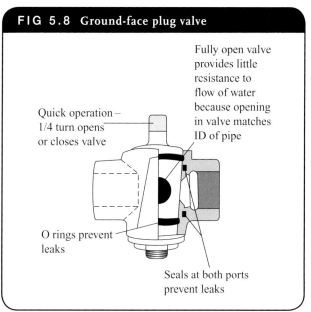

FIG 5.8 Ground-face plug valve

Fully open valve provides little resistance to flow of water because opening in valve matches ID of pipe

Quick operation – 1/4 turn opens or closes valve

O rings prevent leaks

Seals at both ports prevent leaks

FIG 5.9 Ground-face plug tap

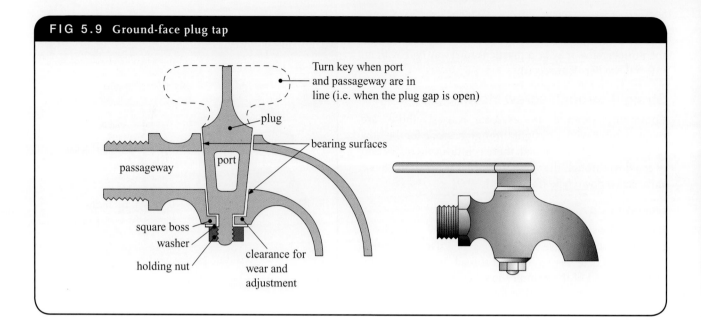

Turn key when port and passageway are in line (i.e. when the plug gap is open)

plug

bearing surfaces

passageway

port

square boss

washer

holding nut

clearance for wear and adjustment

Operation

These taps are simple to operate and require only a quarter turn to open or close the valve. They are manufactured with either T-heads or folding lever handles with provision for locking in the off position (Figure 5.9).

Ball valves

Ball valves are simple and inexpensive valves that are quickly and easily operated (Figure 5.10). Good ball valves allow unrestricted flow when fully open. The valves are used universally throughout the plumbing industry, for hot and cold water and when coded yellow they can be used for gas installation work.

Ball valves contain a ball with a hole through the centre, which is seated within a nylon or similar bushing, and a lever that turns the ball a quarter turn, or 90°, to control the flow. Try to avoid long straight upstream pipework, which may suffer from water hammer when a ball valve is shut off too quickly.

Ceramic disc cartridges

There is little or no maintenance needed for the actual discs compared to the limited life of a traditional jumper valve used in a conventional screw down pattern tap (Figure 5.11).

Single lever mixer—non thermostatic

The mechanism is designed to allow an amount of water from either hot or cold supplies to enter the outlet. You choose the amounts of hot and cold by varying the two ports from the inlets into the one port on the outlet when you operate the single lever on top. As one port closes the other port opens.

Single lever mixers (non thermostatic) have no capability of maintaining constant delivery temperatures when inputs of temperature, pressure and flow rate vary.

FIG 5.10 A typical quarter-turn ball valve

FIG 5.11 A typical quarter-turn cartridge-style tap

SUSTAINABLE PLUMBING: CERAMIC DISC CARTRIDGES

With today's ever-increasing concern for conserving our natural resources, all ceramic disc cartridges with their precise control of water flow provide an ideal valve mechanism for conserving water, as compared to traditional jumper valve systems. Furthermore, many of the components are manufactured from materials that are recyclable.

The quarter-turn cartridge has the main advantage of providing maximum flow rate per degree of rotation due to its butterfly-type aperture. The cartridge offers a compromise between resistance to water hammer, maximum flow rate and low noise levels. (Figure 5.11).

The half-turn cartridge was designed to improve comfort in use by facilitating temperature adjustment through a longer operating run.

The three-quarter-turn cartridge is ideally suited for Australian conditions where extremes of high or low water pressures may be experienced.

Mixing valves (Figure 5.12(a)) are best suited for use where hot and cold water pressures are equalised, such as in a mains pressure storage hot water installation.

If mixing valves are to be used with low-pressure hot water, a special flow control valve must be fitted under the cold water inlet valve seat before the mixing valve will function satisfactorily. There are mixer valves manufactured specifically to perform correctly in low-pressure installations.

Operation

As the name suggests, single lever mixers are taps that operate using one lever to control the water, operating through a replaceable ceramic disc cartridge, as shown in Figure 5.12(b). Generally speaking, you turn the tap to the left to get progressively hotter water or to the right to get progressively colder water. The advantage of a single lever tap is that you can leave the tap lever in your favoured position once you have found a water temperature that works best for you. It is also easy to control the flow of water as the higher you lift the lever, the stronger the flow of water.

Thermostatic control

These fixtures blend hot and cold water before allowing it to flow through a single outlet, as opposed to having two separate fixtures. This enables users to adjust the temperature of the water between the two extremes as they please. There are also a number of safety advantages for both children and adults, as these particular types help to prevent accidents involving scalding, when they are fitted with a thermostatic control.

Float valves

Float-operated valves are designed to regulate the flow of water into a storage tank or cistern. There are several different patterns available, each designed to serve a specific purpose. While they all operate on the same principle, the different patterns are required for larger diameter valves or high pressures. A float valve that operates in reverse is also available. This valve is used on automatic flushing cisterns where an increase in the volume of water is required when the cistern is nearly filled, so that siphonage from the cistern can commence.

FIG 5.12(a) A single lever basin tap

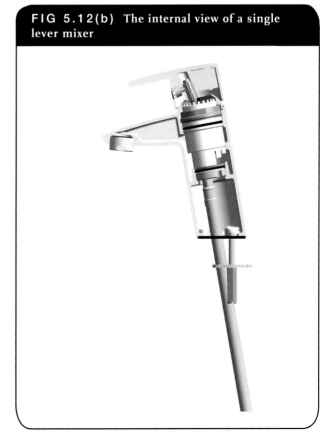

FIG 5.12(b) The internal view of a single lever mixer

Operation

The valve and washer within the float valve are opened or closed by the movement of a cam that is operated by a lever connected to a copper, glass or plastic float. When the cistern is empty, the float hangs down and the valve and washer are held away from its seat by the cam. This allows the water to pass through and fill the cistern or storage tank. As the water level rises in the cistern, the float and lever are raised, which moves the cam and gradually closes the valve. The valve will remain closed until the water level is lowered (Figure 5.13).

Check valves

Check valves are used to prevent back-flow of water. They are of four main types:

1 swing check—allows full flow

2 horizontal lift check

3 vertical lift check

4 spring loaded unicheck valves—horizontal or vertical use.

Swing and horizontal check valves are mainly for a horizontal position (Figures 5.14(a) and (b)).

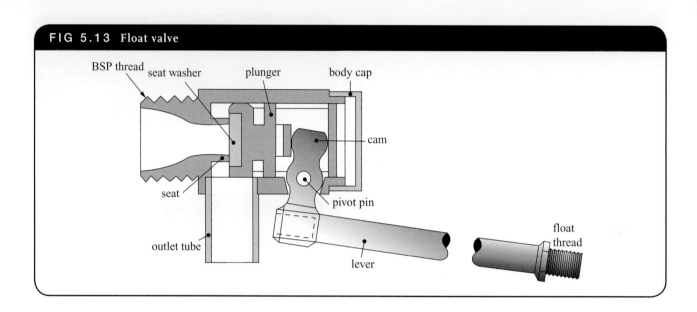

FIG 5.13 Float valve

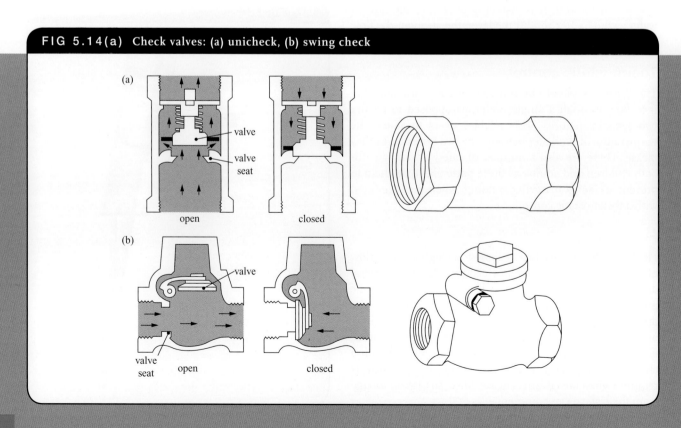

FIG 5.14(a) Check valves: (a) unicheck, (b) swing check

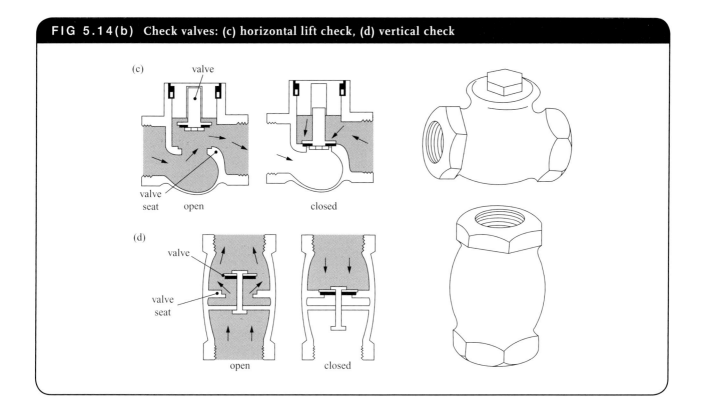

FIG 5.14(b) Check valves: (c) horizontal lift check, (d) vertical check

Operation

Check valves are usually of a simple design in which a loose valve rises to allow flow in one direction but falls back into the closed position when flow stops or back-flow conditions arise.

PRESSURE CONTROL VALVES

Maximum pressure within buildings

As per the AS/NZS 3500.1, the plumber must ensure that the maximum static pressure at any outlet—except a fire service outlet—within a building does not exceed 500 kPa. Pressure control valves can only be used to reduce high inlet pressure. They will not boost low inlet pressure. Pressure control valves may be fitted as controls to a whole service, i.e. hot and cold supply, or be restricted to the hot water system only. They are divided into three groups:

1 ratio or mains proportion valves
2 pressure limiting valves
3 pressure reducing valves.

Ratio or mains proportion valves

Ratio valves are designed to reduce outlet pressure to a set ratio of the inlet pressure, e.g. in a 2:1 ratio valve an inlet pressure of 1000 kPa would give an outlet pressure of 500 kPa. Ratio valves are not suitable for use where excessive pressure fluctuations may occur. For this reason they are not suitable for domestic use as pressure in mains can become low in peak periods. Ratio valves are, however, ideally suited for use as an alternative to break pressure storage tanks in high rise buildings (Figure 5.15).

Operation

Operation relies on pressure acting upon the respective surface areas of the main valve and the smaller bottom piston. For example, if the small piston has half the surface area of the main valve, a 2:1 ratio is produced, or we can say that water pressure acting on opposite surface areas creates a balance of pressure to a set ratio.

Pressure limiting valves

Pressure limiting valves (Figure 5.16) are designed to reduce high inlet pressure to an outlet pressure within acceptable limits. These valves are the most suitable pressure control valves for general domestic use. However, they have a moderate variation over their stated setting. If a set pressure is required a reduction valve should be used.

Operation

When the inlet pressure is less than or equal to the pressure setting of the valve (usually about 500 kPa), the valve remains fully open. Higher inlet pressures act on the small piston and the valve itself, creating a pressure balance that is automatically controlled by the tensioned spring to an acceptable outlet pressure.

Pressure reducing valves

Pressure reducing valves (Figure 5.17) are designed to reduce high inlet pressures to a specific outlet pressure. They would normally only be used when a specific set pressure is required. (Pressure limiting valves are preferred for domestic installations because they create less resistance to water flow and give a greater flow in proportion to size.)

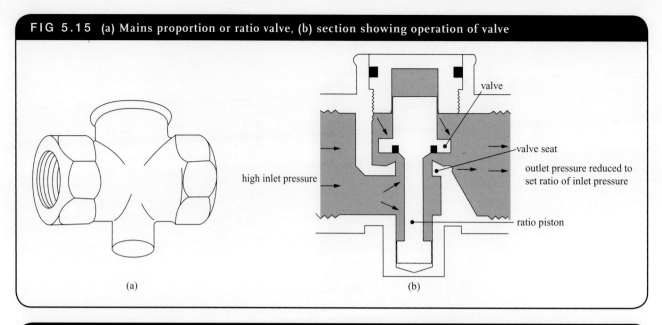

FIG 5.15 (a) Mains proportion or ratio valve, (b) section showing operation of valve

(a)

(b)

valve

valve seat

outlet pressure reduced to set ratio of inlet pressure

ratio piston

high inlet pressure

FIG 5.16 (a) Pressure limiting valve, (b) section showing valve fully open, (c) section showing valve in operation

valve seat

valve

low inlet pressure

same outlet pressure

balance piston

stainless steel spring

high inlet pressure

limited outlet pressure

(a)

(b)

(c)

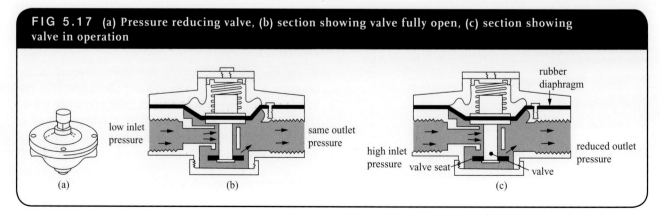

FIG 5.17 (a) Pressure reducing valve, (b) section showing valve fully open, (c) section showing valve in operation

low inlet pressure

same outlet pressure

rubber diaphragm

high inlet pressure

valve seat

reduced outlet pressure

valve

(a)

(b)

(c)

Operation

Pressure reducing valves operate on a similar principle to pressure limiting valves but the use of a diaphragm allows precise pressure control. High inlet pressure lifts the valve closer to the seat (the higher the pressure the closer the valve is to the seat) until a pressure balance occurs. Once a state of balance occurs, the outlet pressure is automatically determined by the tension setting of the pressure spring.

COMBINATION VALVES

Combination valves are used as controls to hot water heaters. They incorporate several different types of valves and can be useful where space will not allow separate fittings—they are normally cheaper than buying separate valves. Variations of combination valves are:

1   combination stopcock, strainer and check valve (Figure 5.18)

2   combination stopcock and strainer

3   combination stopcock and check valve

4   combination strainer and check valve.

Operation

Each section of the valve performs the same task and operates in the same manner as a separate valve of the same type would operate, i.e. the stopcock operates exactly as an ordinary loose valve stopcock would. The strainer is used to filter impurities from the supply and the check valve section prevents back flow of water.

Combination temperature and pressure relief valves

Combination temperature and pressure relief valves (Figure 5.19) are fitted in the top 150 mm or top 20 per cent of mains pressure water heaters. They are fitted to vent hot water to the atmosphere should unsafe conditions arise and so relieve internal stress on the hot water cylinder. Temperature and pressure relief valves also incorporate a drain relief valve and sometimes a vacuum break device.

Operation

Excess pressure is relieved when internal pressure exceeds spring pressure by the valve lifting from its seat until pressure is normalised.

Excessive temperature, which might be caused by a malfunctioning thermostat, causes a temperature rod to push the valve from its seat. The polythene rod has an expansion rate per degree Celsius that is almost ten times that of the outer copper sheath. As the temperature

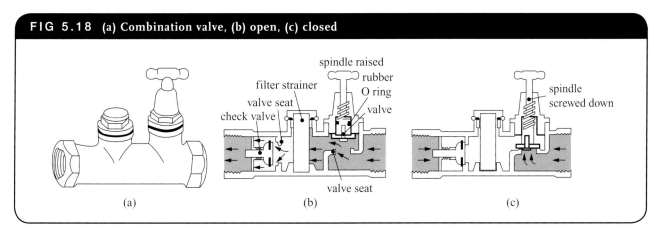

FIG 5.18   (a) Combination valve, (b) open, (c) closed

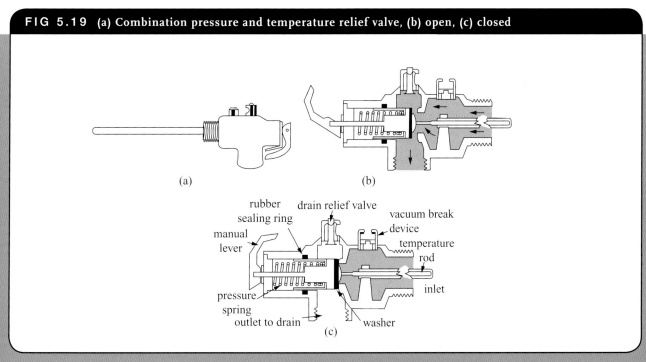

FIG 5.19   (a) Combination pressure and temperature relief valve, (b) open, (c) closed

becomes excessive, the polythene pushes on to the stainless steel pushrod, which in turn pushes the valve from the seat until temperature returns to that required.

The drain relief valve (Figure 5.20) is a further safety precaution and operates only if the drain from the valve outlet has been blocked or damaged. Once the drain relief valve has operated the valve should be replaced. The vacuum breaker device (Figure 5.21) is incorporated to prevent cylinder collapse should a vacuum occur in the hot water cylinder. The valve is held closed by internal pressure and is forced open by atmospheric pressure should a vacuum or negative pressure conditions occur within the cylinder.

VACUUM BREAK VALVES

Vacuum break valves (Figure 5.22) serve two purposes: to prevent backflow of polluted water from subsoil water outlets, such as in lawn sprinkler systems; and to prevent cylinder collapse due to vacuum conditions within a hot water cylinder.

Subsoil outlets may allow backflow under such conditions as a burst water main causing siphonage from the subsoil outlet back to the main opening. A vacuum condition may occur in a hot water heater at any time when water outlets over a fixture are below the heater. In each case the vacuum breaker allows atmospheric air into the pipelines or cylinder, to stop siphonage occurring.

Operation

As with the vacuum breaker device incorporated in temperature and pressure relief valves, the operation is very simple. The valve is held closed on the seat by internal pressure. When vacuum or negative pressure conditions occur, no internal pressure holds the valve on to the seat. External atmospheric pressure forces the valve from the seat and allows air in to stop siphonage occurring. As normal pressure conditions return, the valve is forced back down on to the seat and into a closed position.

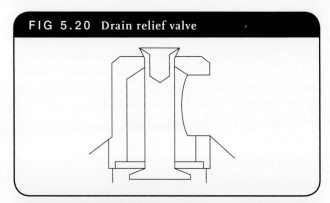

FIG 5.20 Drain relief valve

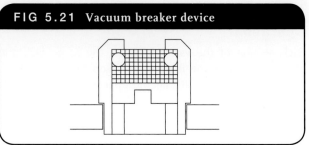

FIG 5.21 Vacuum breaker device

SOLENOID VALVES

Solenoid valves are electrically operated valves that have a wide and varied use in many industrial and commercial installations. For domestic purposes they are usually only used with falling level, roof model storage water heaters, fitted on the cold water inlet but are also used on appliances such as dishwashers and washing machines, as well as for automatic flushing.

Solenoid valves are activated by an electrical impulse sent through electrical mains to a starter switch or by a time clock control. This ensures that the valves only operate during 'off-peak' periods when a cheaper rate of electricity is available.

NOTE: The heating element in falling level storage heaters operates only when the solenoid valve is open and water has filled the tank.

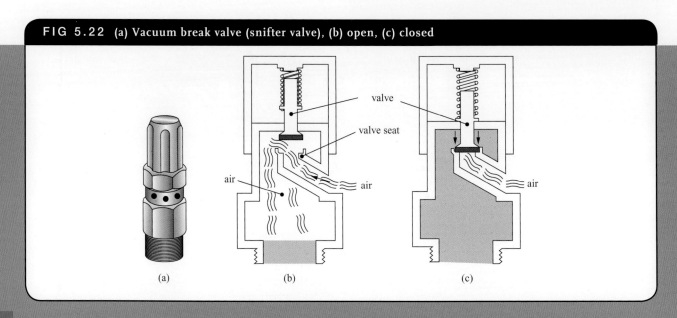

FIG 5.22 (a) Vacuum break valve (snifter valve), (b) open, (c) closed

FIG 5.23 (a) Solenoid valve, (b) open, (c) closed

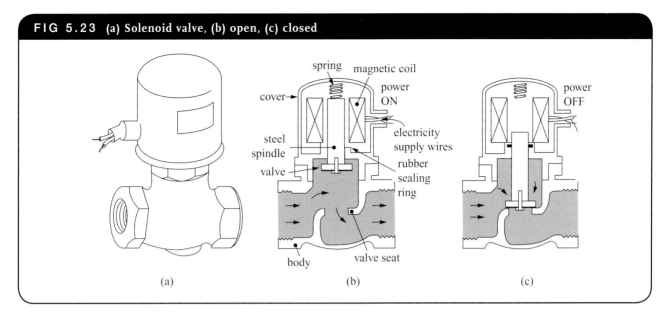

(a)　　　　　　　(b)　　　　　　　(c)

Operation

In the closed position the valve is held against the seat by inlet water pressure. When electricity activates the valve, an electromagnetic coil attracts the valve stem and so lifts the valve from the seat, allowing water to flow through the valve. The valve remains open as long as power is supplied to the electromagnetic coil. When supply of power ceases the valve is again held closed by inlet water pressure (Figure 5.23).

TEMPERING VALVES

An adjustable tempering valve can be installed into the hot water supply line incorporating all household hot water outlets. This will provide tempered water to all hot water fixtures at a temperature that will guard against scalding.

It automatically mixes the hot water with cold water to reach a maximum of 50° C. So in effect any hot water tap has 50° C water coming out of it. It is hot enough to want to mix it with cold water before having a shower. But it will not burn anyone in the short to medium term.

Tempering valves blend hot and cold water to deliver mixed water at a constant temperature. They incorporate a temperature-sensitive element that expands and contracts depending on the temperature of the water flowing across it. This action of the element in turn moves a piston that shuttles between the hot and cold ports. In doing so the valve maintains an essentially constant outlet temperature (50 °C), and thus reduces the risk of scalding accidents (Figure 5.24).

THERMOSTATIC MIXING VALVES

As AS/NZS 3500.4 states,

'All new heated water installations shall, at the outlet of all sanitary fixtures used primarily for personal hygiene purposes, deliver heated water not exceeding:

(a) 45° C for early childhood centres, primary and secondary schools and nursing homes or

FIG 5.24 A sectioned view of a tempering valve

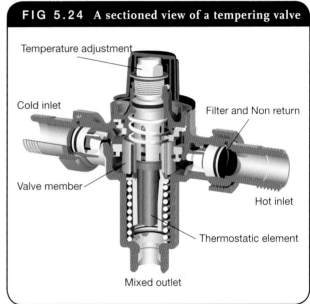

Temperature adjustment

Cold inlet

Filter and Non return

Valve member

Hot inlet

Thermostatic element

Mixed outlet

similar facilities for young, aged, sick or people with disabilities; and

(b) 50° C in all other buildings.'

(Section 1.9.3 (a), (b))

Mixing valves and warm water systems have for many years had application in hospitals and nursing homes. In hospitals, patients' disorientation due to the side effects of certain types of medication or anaesthetic has resulted in many scaldings. Also, many developmentally disabled persons and psychiatric patients are insensitive to the effect of hot water on their skin; scalding may result should they try to adjust normal hot and cold water taps. Authorities and hospital controlling bodies in various states stipulate that hot water for use with ablution facilities must be distributed with cold water and mixed for delivery at the outlets by an approved thermostatically controlled 'thermal shutdown' mixing valve system.

FIG 5.25 A thermostatic mixing valve

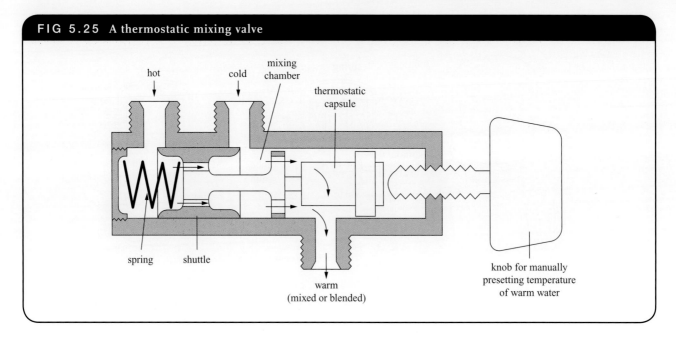

hot · cold · mixing chamber · thermostatic capsule · spring · shuttle · warm (mixed or blended) · knob for manually presetting temperature of warm water

Thermostatic mixing valves also have applications in the home, hairdressing salons, institutions, hostels, dairy farms, chemical and photographic laboratories—in fact anywhere controlled water temperature is desired. They are used for a number of reasons, but mainly because they provide the user with a simplified means of mixing hot and cold water to the desired temperature.

While mechanical mixing valves mix hot and cold water successfully, they are not suitable for all applications. These valves have to be manually positioned to align hot and cold inlet ports to mechanically mix or reach the desired temperature. However, thermostatic mixing valves automatically premix water to the desired temperature by the action of an inbuilt thermostatic element (Figure 5.25). Note that only thermostatic mixing valves are generally provided with 'thermal shutdown' devices and therefore are the only type acceptable for hospital use. Also, mechanical mixing valves are designed mainly for single-fixture installation, whereas thermostatic mixers can be used on single or multiple installations.

Most thermostatic mixing valves available today utilise wax element technology due to its increased performance when compared to the old mechanical technologies.

Thermostatic valve selection

Valve selection and associated pipeline design have to be given careful consideration if the valve is to perform to its design standards. Valve selection for multiple installations requires accurate knowledge of pipesizing and flow requirements.

Some of the factors in choosing a thermostatic mixing valve are:

- Is the valve suitable for the purposes of meeting AS 3500.4?
- Does it comply to the requirements of routine maintenance as per AS 4032.3?
- Is the valve being retrofitted to an existing system?

Velocity

One of the major problems encountered in design is the noise created by flowing water through a small bore pipe. As recommended by AS/NZS 3500.1 the maximum water velocity in piping is to be 3 m/s.

Temperature

Warm water (also called mixed, blended or tepid water) may be described as water that is at a safe temperature satisfactory for showering, bathing or hand washing. AS/NZS 3500 allows for a maximum heated water temperature of 45° C in all healthcare and high-risk applications. However, state health authorities may require a lower maximum than this.

Thermal shut down

Thermal shut down refers to the ability of a thermostatic mixing valve to close automatically if the cold water pressure drops too low or fails completely, thus preventing scalding to the user. The time response varies from valve to valve.

Operation

Hot and cold water enter the valve and mix together in the mixing chamber. The blended water passes through the shuttle ports and surrounds the thermostatic elements.

The thermostatic element reacts to the incoming blended water temperature. Too hot a temperature makes the thermostatic element expand towards the presetting knob. Due to the position of the presetting knob the element is prevented from expanding further and redirects its expansion towards the shuttle, which moves towards the hot water inlet port. As the hot water flow is restricted and the cold water flow is increased, the temperature is reduced to the temperature on the concealed presetting temperature knob.

Conversely, the shuttle moves in the opposite direction if the blended water surrounding the element is too cold.

In the event of the cold water flow being reduced or failing completely, the shuttle (due to the reaction of the thermostatic element) will close off the supply of hot water to the valve. As mentioned previously, this movement (which is termed the 'thermal shut down' protection against scalding) can be as rapid as one-tenth of a second.

Servicing

Servicing should be conducted in accordance with the manufacturer's instructions and AS 4032.3. Today's modern technology and materials, along with extended warranties and cartridge designs, some of which are completely enclosed, have almost eliminated the risky dismantling of precision devices (Figure 5.27). Servicing in the field can only deliver a basic performance assessment when components are replaced. In the field a service technician cannot fully test the performance range of an AS 4032.1 certified thermostatic mixing valve.

Modern valves now have many components manufactured from engineering plastics. Intervention with some cleaning compounds or abrasives should be avoided. Figures 5.26 and 5.27 show a typical valve and valve installation.

The replacement of seals is totally subject to the manufacturer's service directions and limited through AS 4032.3 to a maximum interval of five years. Replacing seals annually when a manufacturer's design does not require this is both time consuming and a waste of funds. Obviously some site conditions may dictate otherwise.

Lubricants must be AS 4020 compliant and suitable for the elastomer, which are recommended by the manufacturer. At all times, the manufacturer should be consulted if there is any doubt.

NOTE: Service and commissioning sheet profiles are contained in AS 4032.3.

When servicing a valve, extreme care should be taken. Although the outer casing of the valve is brass and robust, the inner components are not. It is recommended that each component be removed and placed in a safe place so that it is not likely to fall and be damaged.

Reassembly

If each part is placed in sequence when removed, this will ensure the correct reassembly and help familiarisation with the particular valve. Always use the correct tools; multigrips or Stillson wrenches should *never* be used as they can damage the components considerably.

Replace all seals on service as required. Most manufacturers provide a service kit for their valves. When removing O-rings, do not use an instrument that will gouge or scratch them.

To assist in the reassembly and smooth operation of the valve, lightly smear the threads, seals and component parts with a compound recommended by the manufacturer. Finally, turn on hot and cold water

FIG 5.26 Aquablend 1500 thermostatic mixing valve

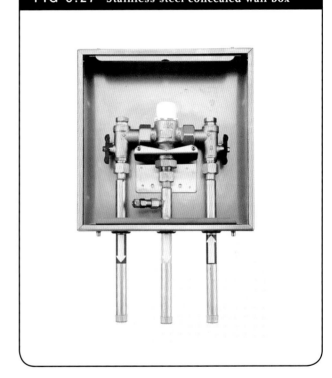

FIG 5.27 Stainless steel concealed wall box

and, with the aid of a thermometer, check that the water temperature corresponds with the preset temperature.

Decontamination

Warm water systems are reticulated water systems that distribute or recirculate warm water through the majority of its branches at a nominal temperature of 45° C by means of a temperature controlling device.

A hot water system reticulates hot water at 60° C through the majority of its branches. The hot water system may include temperature-controlling devices, such as thermostatic mixing valves, to regulate the water temperature to a nominal 45° C to the outlets.

Legionella can multiply at temperatures ranging from 20° C to 45° C (as stated in the SA Health publication, *Guidelines for the Control of Legionella in Manufactured Water Systems in South Australia*).

The disinfection of water systems reduces the numbers of *Legionella*, algae and other bacteria in the systems. Disinfection can be achieved by chlorination (passing a calculated dosage of chlorine through the system) or pasteurisation (allowing hot water at 70 ° C to pass through the system for a period of 5 minutes, or if this cannot be achieved water between 60 ° C and 70 ° C for 10 minutes)

NOTE: When carrying out the decontamination procedure the appropriate PPE must be worn, complying to the occupational health, safety and welfare act in your State or Territory.

Commissioning

When commissioning a new valve or servicing an existing thermostatic mixing valve, controlling authorities require a service log sheet to be submitted. This log sheet should be completed in duplicate and one copy retained by the owner or manager of the property. In some areas the commissioning of new valves and the servicing of existing valves can only be carried out by operators who hold specially endorsed licences.

Changes in technology

TMV monitoring system

Changes in technologies are significant. Additional products, such as a Smartflow TMV monitoring system, allows for 24/7 monitoring of TMV performance in an installation.

The Smartflow TMV Monitoring System is used as a risk-management tool for the performance and maintenance of Aquablend TMVs. The networked system constantly monitors the TMVs electronically, through a calibrated temperature probe installed into the TMV. It provides engineering personnel with a constant temperature readout and performance status report of all TMVs throughout the facility that improves efficiency.

Point of use

The new valve delivers safe warm water in high risk applications such as hospitals and aged care facilities. The self-draining spout design allows heat disinfection directly at the point of use, ensuring there is no bacterial growth, such as Legionella.

FOR STUDENT RESEARCH

1. Investigate and document the authorities that are responsible for the installation and maintenance of thermostatic mixing valves in your state or territory.

2. Review AS 4032.1 (available from Standards Australia, or through your TAFE, school or college library) and research alternative electronic technologies.

Australian Standards

AS/NZS 3500.1 Plumbing and drainage—Water services
AS/NZS 3500.4 Plumbing and drainage—Heated water services

AS 4032.1, AS 4032.2, AS 4032.3 Water supply—Valves for the control of heated water supply temperatures
AS 3718 Water supply—Metal bodied taps specified by performance

REFERENCES

Standards Australia, AS/NZS 3500.4, Section 1.9.3 (a), (b), SAI Global.
SA Health, *Guidelines for the Control of Legionella in Manufactured Water Systems in South Australia*, www.sa.gov.au, 2008.

CHAPTER SIX

Insulation and noise transmission

LEARNING OBJECTIVES

In this chapter you will learn about:

6.1   thermal insulation

6.2   types of insulating materials

6.3   noise transmission in pipework

6.4   how water hammer arrestors work

6.5   water hammer causes and remedies.

INTRODUCTION

Attention must be given in any plumbing system to minimising the heat losses through radiation and conduction caused by exposed surfaces such as hot pipes, storage vessels and boilers. To prevent heat loss some form of heat-resisting insulating material, commonly called 'lagging', is needed. A short definition of 'thermal insulation' is that it is 'a material applied to surfaces in order to reduce the amount of heat emitted by the pipes and boilers'. Heat loss occurs when there is a difference in temperature between exposed pipework and the surrounding air. The efficiency of an insulation material depends upon the number of air spaces existing within the material, its durability and resistance to change by heat.

INSULATION MATERIALS

A large number of insulating materials are in common use throughout the plumbing industry. Each material is designed for specific applications and care should be taken to ensure that the material selected is suitable for both the temperature range expected and the location (Figure 6.1). It is important to remember that some insulating material becomes ineffective when wet, so care should be taken when installing external lagging to protect it from weather and physical damage.

Some common types of insulating materials and their applications are:

1   microcellular PVC (hot and cold copper piping, chilled water)

2   aluminium foil and metal sheathed mineral wood (hot piping)

3   EPDM rubber—ethylene propylene diene monomer (M-class) rubber (exposed solar hot piping, steam lines up to 150 °C)

4   reflective foil laminate and metal sheathed styrene foam (cold piping)

5   reflective foil laminate and metal sheathed isocyanurate foam (cold piping)

6   reflective foil laminate and metal sheathed mineral wool (cold piping)

7   high-density polyethylene-encased and galvanised-steel encased rigid polyurethane foam (hot and cold piping)

8   high-density polyethylene-encased rigid foamed glass (hot and cold piping)

9   reflective foil laminate sheathed foamed glass (hot and cold piping).

SUSTAINABLE PLUMBING: INSULATION SOLUTIONS

The growth of high-density living, residential apartments, hotels, motels, aged care buildings, townhouses and other attached buildings, has resulted in an increased demand for noise abatement methods and materials which will allow for a better quality living environment. The construction industry is now requiring sustainable insulation solutions to protect both the

built environment and the natural environment. The changing regulations and requirements set out by the Building Code of Australia (BCA) regarding energy efficiency and acoustic insulation solutions must be achieved by manufacturing products that meet the highest thermal, acoustic and fire safety performance levels.

FIG 6.1 Insulation to solar close-coupled units needs to be weather resistant and withstand high temperatures (possibly up to 99 °C)

Some other materials that have been used in the past for insulating pipework and that may be encountered during maintenance are:

1 hair felt

2 cork (sheet and granulated)

3 magnesia

4 rock wool

5 asbestos.

NOTE: Care should be exercised. Do not handle asbestos materials as inhaled dust particles may present a health risk.

TRANSMISSION OF NOISE IN HOT AND COLD WATER SYSTEMS

Sanitary fittings associated with hot and cold water systems need to be conveniently placed in any building but should not be obtrusive. Water services are potential sources of noise and the planning and installation of the services should therefore be carefully carried out to avoid the production and transmission of noise.

Sound is louder and travels much faster in water than in air; a noise transmitted through the air may be inaudible from a certain distance but quite audible if transmitted the same distance through a water service. Noises in water services, however, vary in kind and intensity. Some noises are caused by:

- water being disturbed as it flows through pipes and fittings
- peculiarities in the moving parts of taps and valves
- the behaviour of water at the discharge point.

Disturbance through pipes

Noise can be transmitted by the pipes themselves in larger residential buildings such as home units (Figure 6.2). Plumbers may signal to one another by tapping on the pipe

FIG 6.2 The BCA requires waste pipes to be acoustically insulated, especially where they pass over habitable areas

with a metal tool. Probably the most common and obvious noise in water services is the 'singing' or 'humming' noise caused as water passes through the pipe. All internal pipe surfaces tend to cause friction, which forms eddies in the flow, and it is the eddying of the water that causes this noise. The noise can also be intensified by a number of factors such as roughness of the internal pipe surface, a large number of joints, an acute change of direction and the velocity of the flow (Figures 6.3 and 6.4).

FIG 6.3 Joint in drawn copper tube: (a) disturbed flow, (b) streamlined flow

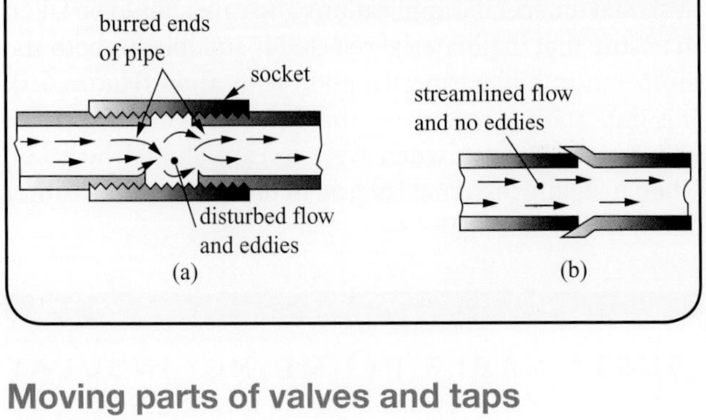

burred ends of pipe

socket

streamlined flow and no eddies

disturbed flow and eddies

(a)

(b)

Moving parts of valves and taps

High-pitched humming or screaming noises can be caused by a jumper valve spinning in a screwed-down valve or tap. This is caused by the water passing over the serrated face and edges of a worn tap washer. The loose stem of the jumper valve within the spindle produces the noise. It can usually be cured by replacing the worn washer.

Chattering noises can also be caused by loose parts such as jumpers in screwed-down taps and valves, and lever

REDUCING NOISE IN PIPES

The following precautions should be taken to reduce or prevent the cause of friction noise in pipelines.

1. Use pipe that has a smooth internal surface. The internal mirror finish of drawn copper pipe is preferred to that of the rougher internal finish of galvanised mild steel.

2. Unnecessary joints should be avoided—plan the job to overcome this problem. Ream any cut ends of tube with the appropriate tool to allow a full bore and smooth end. Where a fitting is necessary to make a joint, give preference to a fitting that is the same bore as that of the tube to which it is to be fitted.

3. Any changes in direction should be made by bend if practicable. It is generally accepted that if the radius of a bend is not less than five times the diameter of the pipe, the velocity of the flow will not be reduced. The smaller the reduction in velocity, the smaller will be the resulting noise-causing eddies.

4. It is essential to consider the velocity of the flow in planning a pipeline; all the pipes should be large enough to ensure that when water is drawn off, the rate of flow is within the capacity of the pipe. Unfortunately, this is not always given sufficient consideration by many plumbers who may, in picking up an individual fitting that is too small, dramatically reduce the pipe size in the interests of economy.

FIG 6.4 (a) Disturbed flow in elbow, (b) streamlined flow in bend

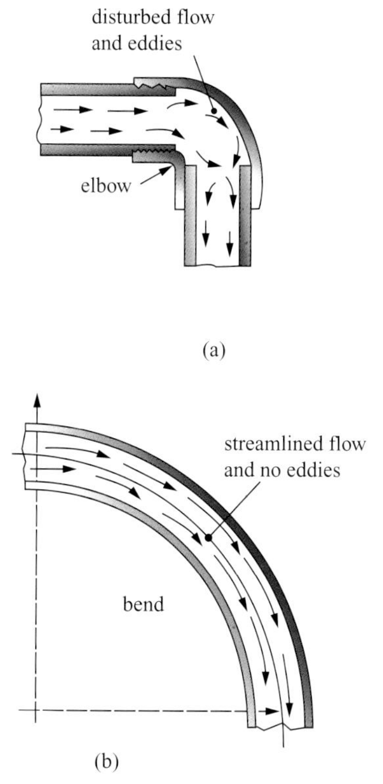

arms and pistons in ball valves; parts of the taps and valves become worn causing them to loosen. To remedy this situation either the worn parts are built up by the addition of new metal or the offending part or valve is replaced.

Behaviour of water at the discharge point

A common noise in water pipelines is the screaming caused by water flowing through a restricted orifice or opening smaller than the bore of the pipe through which the bulk of the water is flowing. The greater the pressure, the greater will be the flow through the orifice in a given time, which increases the noise.

Ball float valves, because of their design, are a common cause of this noise. In most ball float valves the size of the valve orifice is less than that of the pipe supplying the valve; the shape of the piston chamber and ball float valve outlet may also restrict the flow. The noise frequently increases in volume before the valve orifice is closed, when the flow of water is restricted by the face of the piston as it approaches the valve orifice. To eliminate or reduce this noise, close the stop valve controlling the ball float valve sufficiently to reduce the noise. At the same time allow sufficient water to flow through to the ball float valve at a reasonable rate.

WATER HAMMER

Water hammer may be defined as a noise resembling a succession of hammer-like blows on a water service pipe. It is a shock wave that may be felt and heard. For all practical purposes water may be considered to be incompressible when travelling at high speed. If suddenly arrested, by the closing of a valve or tap, a great force is exerted; the pressure is increased above the flow pressure by approximately 350 Pa for each 300 mm/s of sudden decrease in water flow. For example, if a water flow of 3 metres is suddenly arrested, it produces a maximum water hammer pressure of 3500 Pa or 3.5 kPa. The principle of the hydraulic ram is based on the action of water hammer.

Water hammer occurs frequently; the shock may be transmitted throughout a water service and through the main supply to other services such as a hot water service. The high pressures that develop may cause damage because the increase in pressure results in tension stresses being increased on both the pipes and fittings, which may burst as a result. The added strain from this increase is put on all valves and taps in the service. A decreasing 'pulsating' effect may also be set up in the water; this can be felt and heard, and may become a nuisance to the occupiers of the premises.

The intensity of water hammer is influenced by three main factors:

1 Loose pipes causing a drumming effect as they vibrate against the walls or foundations, especially a wooden bearer foundation or metal-clad wall.

2 Closing the tap or valve too fast. This increases the intensity of the hammer.

3 Pressure in the service. The higher the pressure in the service, the more severe will be the hammer.

Locating a water hammer

The most common sources of water hammer are:

1. Spring-loaded taps. These are not commonly used on most domestic services, but could be installed, for example, as a bubbler for children. These snap close unless they are of the non-concussive type.

2. Rubber washers on taps and valves of the hemispherical type. These are usually composed of rubber, sold with the jumper part for the repair of a screw-down tap or valve. This type of washer not only closes the orifice suddenly but also presses down into the tap or valve orifice, transmitting additional thrust to the water in the service after the orifice is closed.

3. Ball float valves. These are not a major cause of water hammer, because they normally close slowly, but they may set up a secondary hammer after the valve has been forced open by the initial hammer and the ball-float has been caused to bounce.

4. Plug cocks or low-pressure taps. These are not normally used on high-pressure services, because there is no method of preventing back feed into the main supply, and this is prohibited by most statutory authorities. These plug cocks or low-pressure taps, however, are sometimes installed inadvertently by the owners of premises, and the sudden stop to the flow of water causes water hammer.

5. Sharp changes in direction, which can set up **turbulence in flow.** When the water flow on the outlet side is stopped suddenly, the water entering the inlet side of the bend hits against the water that is stationary.

6. Incorrect clipping or support for pipework

Removal of a persistent water hammer

Should the remedies recommended above or the cost of rearranging the installation be too expensive or unwarranted, a water hammer arrestor should be installed as near to the tap or valve causing the water hammer as practicable. To understand the principles of water hammer arrestors, we can look at how an air chamber works.

REMEDIES FOR WATER HAMMER

1. Fit spring-loaded taps with a flow-control valve, which is a non-concussive spring-loaded valve designed to close slowly to avoid causing water hammer.

2. Replace unsuitable tap washers with flat fibre or neoprene washers.

3. Decrease the supply of water to ball float valves

4. Replace low-pressure taps with high-pressure taps of the loose-valve type, to bring the service into line with the regulations (most statutory authorities prohibit the use of plug cocks on high-pressure services).

5. Rearrange the water service layout to replace sharp changes of direction—it may be expensive, but necessary. Use easy radius bends on all water services to reduce the risk of water hammer.

6. Increase the number or type of pipe supports.

7. Ensure accurate pipe sizing to limit the velocity to 3 metres/second.

8. Maintain a maximum of 500 kPa pressure.

Principles of an air chamber

The air chamber needs to be strongly constructed—standard pipe fittings are often used and are perfectly airtight. Provision must also be made for draining the air vessel for the purpose of replacing the air that becomes absorbed by water.

When the bulk of the air in the chamber is absorbed the water will be heard again; when this stage is reached it is necessary to replenish the air in the chamber. The air may be renewed by draining the service of water and allowing air to enter through an adjacent valve or a special valve in the vessel itself. This valve needs to be in an accessible position or the renewal of the air may be neglected. Figure 6.5 shows various types of air vessels.

The problem of noise or water hammer in water services is not a simple one. A noise may be caused in one

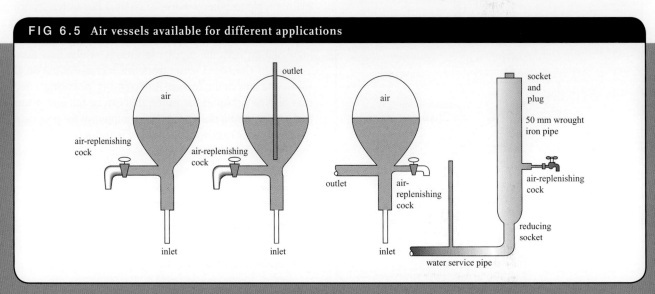

FIG 6.5 Air vessels available for different applications

FIG 6.6 (a) and (b) Examples of modern water hammer arrestors

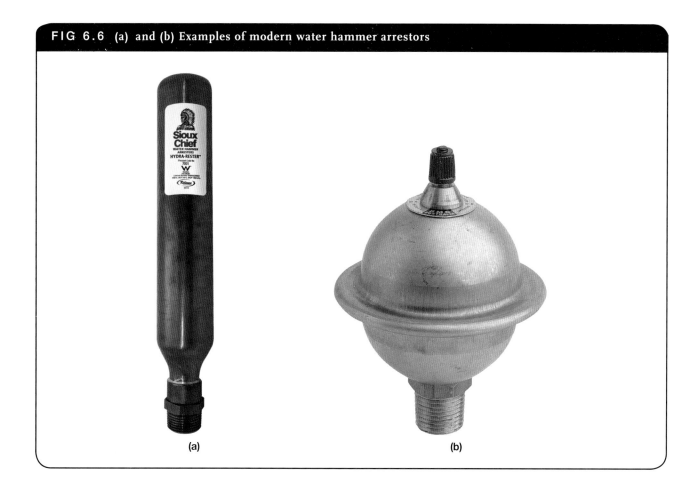

(a)   (b)

place and yet be transmitted a considerable distance; the noise may not necessarily be caused by the valve or tap that is first opened. Finding the source of the noise or hammer can be time consuming and thus expensive.

Modern water hammer arrestor

Modern water hammer arrestors work on the principle of compressing air to absorb the induced pressure wave. The arrestor is placed at 90 degrees to the water flow, just upstream from the offending device. When the device closes, the pressure wave begins to move back along the pipe. As it reaches the arrestor the force acts against a piston, which in turn pushes against a charge of air that is compressed, dissipating the wave energy and nullifying the hammer effects (Figure 6.6).

FOR STUDENT RESEARCH

1. With the assistance of your supervisor, investigate and document any signs of water hammer within your school campus.

2. Investigate and document the approved methods of disposal of asbestos products in your area. List the state or local authorities responsible for these.

Australian Standards

AS/NZS 3500.1 Plumbing and drainage–Water services

AS/NZS 3500.4 Plumbing and drainage–Heated water services

AS 1530.3 Methods for fire tests on building materials, components and structures

AS/NZS 4859.1 2002 Materials for the thermal insulation of buildings

AS/NZS 2712 Solar and heat pump water heaters–Design and construction

7 Hot water systems

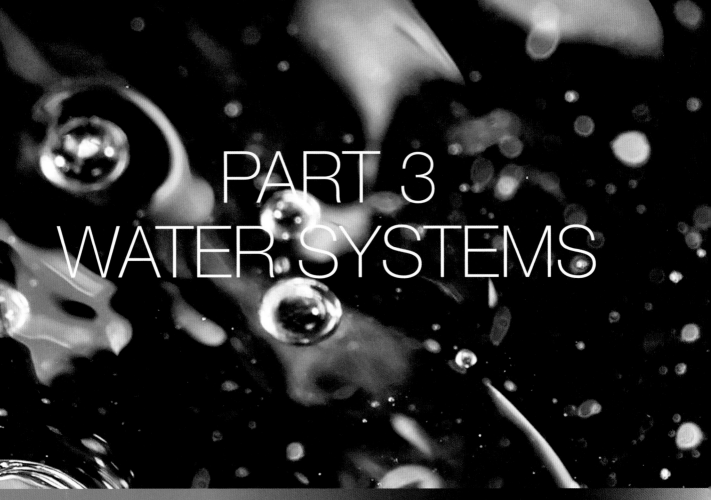

PART 3
WATER SYSTEMS

Plumbing Services Series

CHAPTER SEVEN

Hot water systems

LEARNING OBJECTIVES

In this chapter you will learn about:

7.1 **the principles of heat and heat transfer**

7.2 **how these principles relate to water heaters**

7.3 **different types of water heaters available**

7.4 **hydrostatic testing**

7.5 **commissioning of hot and cold systems.**

SOME DEFINITIONS

Heat
A form of energy capable of performing work.

Sensible heat
Heat which can be sensed by touch or feel. It is the heat taken in or given out when the temperature of a body is raised or lowered and the intensity/amount of sensible heat is reflected by the water's temperature. The higher the temperature the more severe the effect of scalding will be.

Latent heat
When a body passes from a solid to a liquid state or from a liquid to a gaseous state, it absorbs a certain amount of heat energy without changing its own temperature; the heat absorbed is 'latent' or 'hidden' and is not indicated by a thermometer.

The amount of energy needed to change the state of a substance is approximately four times the amount of initial energy needed to raise the substance's temperature from one state to another. It is this hidden heat that makes steam a good source of energy as well as making it extremely dangerous: steam burns deep.

Specific heat
As different substances vary in structure and make-up, they require different amounts of heat to raise an equal amount (or mass) of that substance through the same temperature. Specific heat may be defined as the amount of heat required to raise the temperature of one kilogram of a substance by one degree Celsius.

INTRODUCTION

Hot water systems are often referred to as hot water heaters, but this is a misnomer because hot water does not need heating—it is already hot. Therefore, in this chapter we will refer to these devices as either hot water systems or simply as water heaters.

PRINCIPLES OF HOT WATER PLUMBING

Some of the terms used in hot water plumbing are important and must be understood before considering the practice of hot water plumbing.

Measurement of heat energy

A unit of heat is based upon its effect in increasing the temperature of water. The units of heat used are as follows:

1 **Joules** The amount of heat energy (specific heat) required to raise one litre of water through one degree Celsius is 4186 joules.

2 **Kilojoules** One kilojoule equals one thousand joules (1 kJ = 1000 J).

3 **Megajoules** One megajoule equals 1000 kilojoules (1 MJ = 1000 kJ = 1 000 000 J).

To heat one litre of water through one degree Celsius, 4.2 (4.186) kJ or 0.0042 MJ of heat energy is required. As gas heaters are rated in MJ, 0.0042 is often used to calculate heat energy usage for water heaters.

For example, to calculate the energy required to raise 130 litres of water from 15° C to 65° C, the following calculation would be used:

```
HE = L × T × C
Where HE = Heat energy required in MJ
        L = Water heater capacity in litres
        T = Temperature rise in degrees Celsius
        C = Coefficient or specific heat of water
Therefore:
        HE = L × T × C
        HE = 130 × 50 × 0.0042
        HE = 27.3 MJ
```

If a storage water heater is rated at 30MJ/hr then it will take just under one hour to raise the 130 litres of water from 15° C to 65° C, not allowing for losses in efficiency.

To change the state of water to steam requires much more energy. To raise one litre from 0° C to 100° C requires:

```
HE = L × T × C
HE = 1 × 100 × 0.0042
HE = 0.42 MJ
```

But to change it into steam at the same temperature requires approximately another 1.6 MJ per kg. So the total energy required to raise one litre of water from 0° C to 100° C and turn it into steam requires an initial 0.42 MJ to raise its temperature, plus an additional 1.68 MJ to change its state from liquid to gas which equals 2.1 MJ per kg. Therefore one kg of steam at 100° C contains approximately five times the amount of energy than the water it was produced from. If water is allowed to turn into steam, it occupies 1600 times its original space and in a closed vessel such as a water heater this can have a catastrophic effect. This is why we have safety devices to prevent this from occurring.

Another reality is that the water in the water heater will expand as it heats and this is allowed for by either an expansion allowance (e.g. expansion tank) or pressure relief valve. The amount of expansion can be calculated by using the following formula:

```
E = L × T × C
Where E = Expansion measured in litres
        L = Water heater capacity in litres
        T = Temperature rise in degrees Celsius
        C = Coefficient of expansion for water (0.000375)
Therefore:
        E = L × T × C
        E = 130 × 50 × 0.000375
        E = 2.4375 litres
```

Heat transmission

Heat may be distributed or transmitted in three ways—conduction, convection and radiation.

Conduction

Conduction is the transfer of heat from particle to particle of a substance or from one body to another by physical contact of these bodies. Conduction will only occur when the particles or bodies are at different temperatures, with the heat flowing from the hotter body to the colder until both are at the same temperature. Therefore, while a fire is burning in a boiler, heat is transmitted through the particles of metal forming the shell until the interior becomes hot, heating the water contained in the shell. Most metals are good conductors of heat, the best being silver, gold and copper. The poorest conductors are tin, lead, platinum and bismuth.

Convection

Convection is the transfer of heat through a fluid by the circulation of that fluid over a hotter or colder body. When heat is applied to the bottom of a hot water cylinder, the particles of water in contact with the heated base of the cylinder will be heated and expand, becoming lighter than the colder particles of water in the upper part of the cylinder. Because of this the hot water will be displaced and forced upwards to the top of the cylinder by the heavier particles tending to sink and take their place. These in turn become heated and are forced upwards by other particles seeking to take their place, setting up a circulatory movement termed 'convection', which will continue as long as the heat continues to be applied.

Radiation

Radiation is the transmission or diffusion of heat by means of heat rays that are given out in straight lines in every direction from the source of the heat in the same way that light is given out from a lamp. These rays pass through the air without heating it; no heat is experienced from radiant heat until the heat rays strike or are intercepted by a solid body.

An example of how radiation works will help to explain the concept. A person sitting in front of an electric radiator feels a sensation of warmth. If a screen is placed between the person and the radiator this sensation of warmth stops immediately, demonstrating that the air surrounding the person has not received heat due to radiation.

When radiant heat encounters an obstacle and heats it, that obstacle becomes a heat source and in turn transfers its heat by conduction or convection and, depending on its surface, may radiate it away again. The surface condition of a body plays an important part in the proportion of heat absorbed or reflected. A rough, dull surface will absorb most of the heat and reflect very little, whereas a highly polished surface will reflect a large proportion of heat and absorb only a small amount. This principle will be examined more fully in the section on solar heating. Figure 7.1 demonstrates the three forms of energy transfer.

HOT WATER SUPPLY

The provision of hot water for domestic use has a long history. The Romans developed sophisticated plumbing systems for private houses and villas and for public baths in the first century BC. The availability of hot water in abundance and at relatively cheap cost commonly marks an important stage in the development of an industrial society. Like many of the services that we now take for granted in everyday living, the provision of hot water requires a great deal of careful planning at several levels. Hot water systems are complex and they will be dealt with here in some detail.

The selection of a particular hot water system will remain finally a matter of personal choice. However, a number of factors must be taken into account before that choice is made. The energy supply or fuel available, and practical considerations such as the amount of water

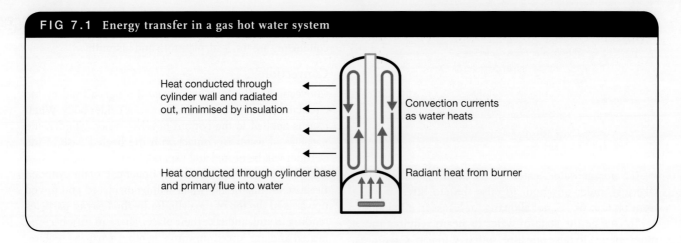

FIG 7.1 Energy transfer in a gas hot water system

Heat conducted through cylinder wall and radiated out, minimised by insulation

Convection currents as water heats

Heat conducted through cylinder base and primary flue into water

Radiant heat from burner

required, the size of the heater and the overall appearance of the system, will all help to determine which system is most appropriate.

Heating fuels

The following heating fuels are the most commonly used throughout Australia for domestic water heating systems:

1 gas

2 electricity

3 thermal energy

4 solar energy

5 liquid fuels

6 solid fuels.

These fuels are used in conjunction with a large selection of different types of hot water systems.

Gas

There are two main types of gas used for heating water: natural gas and liquefied petroleum gas (LPG). Natural gas appliances operate at lower pressures than LPG appliances. Appliances designed to operate on natural gas therefore require modification for LPG and vice versa. Gas tariffs vary considerably but there is usually a block tariff calculated according to the amount of gas consumed. The average cost per unit of gas is generally lower than a comparable quantity of electricity. The efficiency of heat transfer in a domestic gas water heater, however, is less than that of an electric heater due to heat losses up the flue.

When the cost of a unit of gas and the efficiency of it are taken into account it will be found that gas charges are of the same order as electricity rates. Gas, on the other hand, has a lower carbon footprint. Older gas water heaters used to have a lot of their energy 'going up the flue' but newer products recirculate the flue products to harness this previously wasted energy.

Electricity

A minimum charge is levied on all households even if no electricity is used. This is often referred to as an availability charge. In most states of Australia, three flat rates are charged depending on the amount of electricity used. These flat rates are:

- controlled continuous rates
- off-peak rates
- extended off-peak rates.

Off-peak rates

These rates are about half the cost of the lowest continuous rates. They are particularly attractive for storage hot water systems.

The off-peak systems heat water during the hours when most people are asleep, contain and store it ready for use the next day. Because such systems operate outside peak demand periods, off-peak electricity is charged at a lower tariff than systems that use electricity continuously. Most manufacturers offer units with booster elements that allow heating during peak hours if your demand is unusually heavy. The cost for a booster operation is at normal electricity rates.

Depending upon legislation and electricity provider requirements, off-peak rates may not be available unless coupled with energy-saving devices, such as solar panels. In these situations, extended off-peak may be the only available alternative to continuous rates. Always check with authorities and providers before selecting a water heater.

Electricity sourced from renewable sources such as wind farms or hydroelectricity has a low carbon footprint, but most electricity is still produced in coal-powered generators.

Extended off-peak rates

These are available in most states. They are higher than normal off-peak rates but lower than continuous rates and they give a longer heating period than off-peak. These rates may be utilised to boost solar systems to keep costs as low as possible.

Smart rates

The new 'smart meters', which record the time at which the electricity is used, charge varying rates according to time of usage. Heaters connected to this billing system can be connected to a timer to take advantage of the lower rates and avoid peak times.

Thermal energy

Heat pumps take advantage of the heat energy in the surrounding atmosphere. They utilise the refrigeration principle to transfer this energy from the atmosphere to the water (see Figure 7.8(a) on page 94). They still require a small amount of electricity to operate the compressor and an auxiliary form of energy to boost water temperature when needed.

Solar energy

These systems have greatly increased in popularity in recent years, and are available and in use in all states. Because their location affects the economics of the solar water heating unit, the additional cost of a booster system may be necessary in some areas.

Liquid fuels

There is little demand for liquid-fuel-operated domestic water heaters because any savings in fuel costs are more than offset by their low overall efficiency, inconvenience of handling and the need for storing bulk supplies of heating oil, diesel fuel, distillate or kerosene. They may still be in use in some country areas.

Solid fuel

Solid fuel such as wood or coal can also be utilised to provide energy to heat water. These are usually supplementary methods such as a wetback in a slow combustion heater or stove. Wood is the usual fuel in these appliances but coal can also be used.

HOT WATER SYSTEMS

Hot water systems may be divided into two different types: non-storage and storage.

Non-storage heaters

Non-storage water heaters are designed to operate whenever a hot tap served by the heater is opened, and to continue to operate for as long as hot water is required. The automatic operations performed by instantaneous and continuous flow water heaters necessitate the provision of a number of moving parts. The mechanisms of these types of heaters are quite complex.

Instantaneous heaters

Instantaneous heaters normally operate directly from the domestic cold water service at mains pressure (Figure 7.2). Since they are not required to store water, the heaters are small and can be installed in convenient places—under sinks, for example, which can be an important space-saving option in smaller homes and units.

INSTANTANEOUS HEATERS

Advantages:

- cheap to install
- economical for light use (small domestic requirements in flats and units)
- compact
- provides hot water on demand.

Disadvantages:

- if electric, requires three-phase wiring
- water temperature varies with flow
- difficulty is experienced in regulating tap due to changes in flow
- not economical for heavy use
- will supply only one tap at a time.

Continuous flow water heaters

The mechanisms of this type of heater are controlled by a sophisticated computerised panel. The burner of a continuous flow water heater is designed to vary with the flow of water, thereby keeping the temperature of the water constant irrespective of the flow. The heater works at varying gas pressures directly related to the amount of gas required by the burner to provide the temperature required at the selected flow rate.

Continuous flow water heaters (Figure 7.3) normally operate directly from the domestic cold water service at mains pressure. Low-pressure models are available for poor supply or pump pressures (check with the manufacturer as to suitability). Since they are not required to store water, the heaters are small and can be installed in convenient places such as on balconies of units or down the side passageway of a house, saving space in small homes and units.

Storage heaters

Storage heaters heat and store water in an insulated tank until the water is needed. Most domestic storage heaters range in size from 50-litre to 400-litre capacity (and may be larger for commercial storage applications). The quantity of hot water to be heated and stored must satisfy the demand in relation to the hours of the day or

SUSTAINABLE PLUMBING: ENERGY-EFFICIENT WATER HEATERS

In order to meet sustainability targets in building design, high-efficiency heaters are to be used, such as natural gas, heat pumps and solar powered units. Harnessing the power of the sun not only lowers your energy bill but also conserves our precious resources.

There are many energy-efficient water heaters available and you should research the alternatives available to match the needs of your client and available resources.

FIG 7.2 (a) Instantaneous gas water heater, (b) electric instantaneous water heater

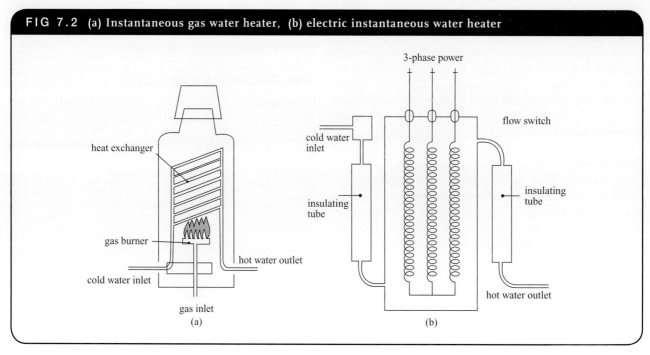

FIG 7.3 Continuous flow water heater

CONTINUOUS FLOW HEATERS

Advantages:

- unlike instantaneous water heaters, can serve more than one tap at a time at the predetermined temperature
- cheap to install
- economical for light use (small domestic requirements in flats and units)
- compact
- provides hot water on demand
- environmentally friendly.

Disadvantages:

- requires a larger gas supply pipework than storage heater due to high energy requirement
- wastes water while the system purges itself
- requires the use of electricity in some cases to power the inbuilt computer.

night during which the hot water may be replenished. The stored water should be maintained to a minimum temperature of 60° C to prevent the build-up of Legionella.

Frequently, it is necessary to store the required water for periods of up to one day. When advantage is taken of a favourable off-peak tariff, water is heated during a restricted period, usually between late evening and early morning. In some areas this is extended into the daylight hours but not during peak demand hours. Often the off-peak option is only available when replacing an existing off-peak hot water system with the same type (i.e. like-for-like replacement) as these systems are considered energy

intensive and less intensive systems are recommended to be used instead.

When electricity can be utilised continuously to heat water, the storage capacity is usually no more than 25 to 80 litres (quick recovery), as this method of heating water is also energy intensive and an expensive option. However, it is often the only option in shops, small units or apartments.

Mains pressure heaters

Mains pressure heaters enable hot water to be delivered at mains pressure and allow advantage to be taken of storage, water heating tariff (off-peak or extended off-peak) and smaller hot water pipes. The level at which the tank

FIG 7.4(a) Mains pressure heater (quick recovery): electric

FIG 7.4(b) Mains pressure heater: gas

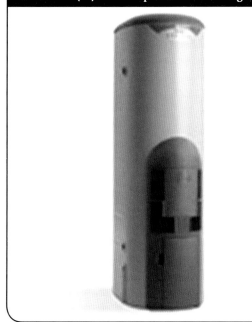

MAINS PRESSURE HEATERS

Advantages:

- easy to install
- easy and cheap to service
- more than one outlet can be used
- high pressure can be utilised for dishwashers and washing machines
- requires minimal space
- quick recovery of water if using continuous rating.

Disadvantages:

- considered a large energy user and may not be able to be installed in new dwellings
- once water is used, need to wait until next heating cycle when using off-peak electricity tariffs.

elevated (Figure 7.5). Flow is by gravity and an exhaust or vent pipe is fitted. These heaters were very popular in rural areas in houses with small roofs where roof models could not be installed. They can be connected to a rainwater tank of not less than 10 metres in height for heaters that were made from Cusilman bronze (no longer available).

The valved version (Figures 7.6 (a) and (b)) is connected directly to the mains supply through reducing valves similar in appearance to a gas regulator operating at pressures of up to 140 kPa. Although no longer available, there may be some still in use due to their longevity. When the valves need replacing, you can no longer source new valves (although reconditioned valves are available).

FIG 7.5 Cistern-fed medium pressure heater

cold feed

vent

minimum 10 m

thermostat

element

is installed is not critical, so these heaters are suitable for floor mounting. These heaters are designed with a cylinder to withstand pressures exceeding mains water pressure, and utilises either electricity or gas as their energy source (Figures 7.4 (a) and (b)).

Medium-pressure heaters

There are two types of heaters included in this category: the 'cistern-fed' model and the 'valve-controlled' flow (controlled mains pressure). Cistern-fed heaters were the original floor type heaters and are a displacement unit with a strong inner cylinder to allow the cistern to be

MEDIUM-PRESSURE HEATERS

Advantages:

- ideal for homes with rainwater tank supply
- can be used on off-peak or continuous rating electricity supply
- can be used on:
  - (a) mains pressure
  - (b) cistern feed in ceiling
  - (c) with a reducing valve and relief valve
- long–lasting.

Disadvantages:

- poor flow due to reduced pressure
- valves can be difficult to obtain.

The cistern-fed models are still produced by specialist water heater manufacturers. The cylinders are made from copper and have a maximum operating head of 7 metres, which is just under 70 kPa. Mains pressure type heaters can be used with this piping set and the maximum head is then only limited by the maximum operating pressure the heater can withstand.

Electric heat-exchange model

Heat is transferred by means of a heat-exchange operation in which water at mains pressure is carried in a coiled copper tube through a body of water that is heated but never used.

As the cold water flows through the coil, it draws heat from the stationary water (heat exchange). A restrictor is used to control the flow of water to enable the maximum heat to be used from the heated body of water (Figure 7.6(c)).

ELECTRIC HEAT-EXCHANGE MODEL

Advantages:

- mains pressure operation
- long-lasting copper construction.

Disadvantages:

- flow poor because of restriction
- heat transfer insufficient to allow large draw-offs
- stationary water requires constant filling in the cylinder as water is lost through evaporation.

Low-pressure types

Roof models of the displacement type are named thus because the cold water displaces the hot water. Earlier models had the cold water feed tank or cistern supply on the top of the water heater (Figure 7.7 (a)). The flow of water is regulated by means of a ball valve and then directed to the bottom of the heater. The draw-off of hot water is at the top and very little mixing occurs because of the tendency of hot water to rise and sit on top of the cold water. This is called stratification and occurs in all storage-type heaters.

FIG 7.6(a) Valved, controlled flow or controlled mains pressure heater, **(b)** medium pressure heater without relief valve

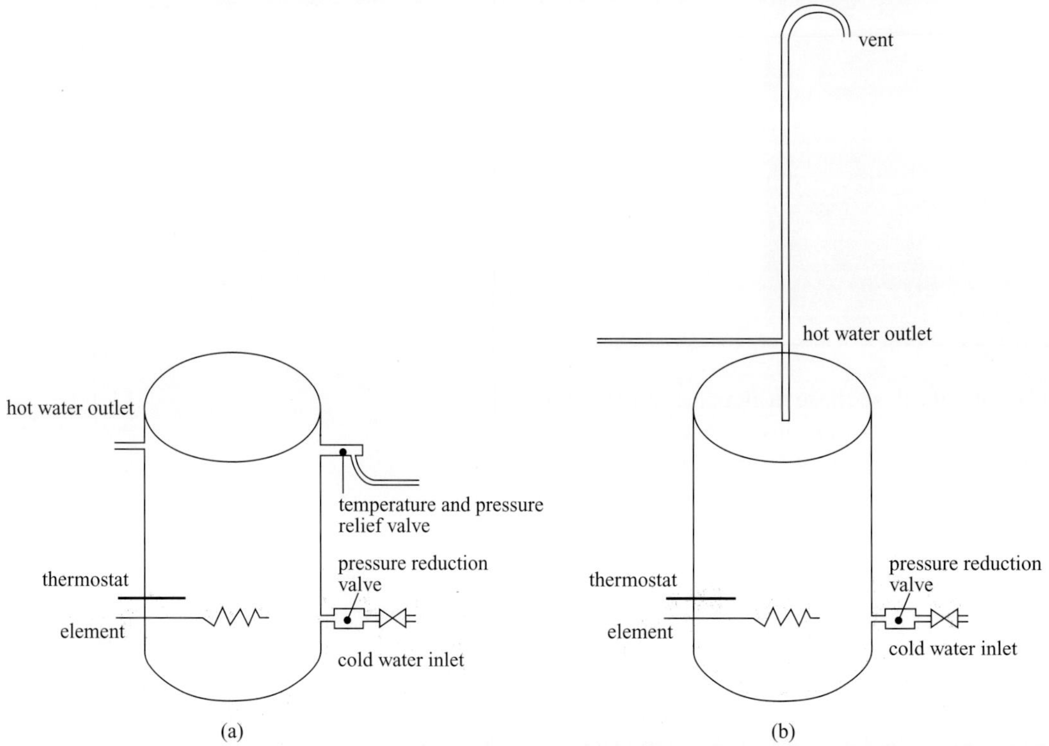

(a)   (b)

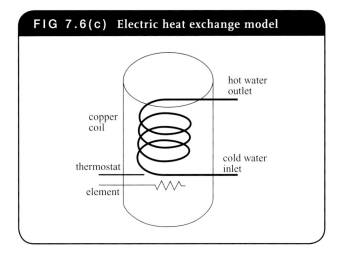

FIG 7.6(c) Electric heat exchange model

FIG 7.7(a) Displacement-type heater (cylinder can be mounted in the ceiling or on the floor with a maximum head of 7 to 7.5 metres)

Side-feed displacement models

These are similar in operation to the displacement roof model but with the cold water feed or cistern lowered and attached to the side of the water heater and hot water drawn off at a point below the water level. This design has developed because it allows lower height and removes pressure from the inner cylinder, which should ensure longer life (Figure 7.7 (b)).

SIDE-FEED DISPLACEMENT MODEL

Advantages:

- longest lasting
- minimal operating parts, trouble-free
- easy to service
- occupies no floor space.

Disadvantages:

- low flow and pressure

FIG 7.7(b) Side-feed displacement heater

The biggest disadvantage of this system is that it will not supply more than two outlets at once.

Heat pump

Heat pumps draw their energy from the surrounding air (thermal energy) to heat water stored in a cylinder. They use the refrigeration principle (Figure 7.8(a)) to draw the heat out of the air and transfer it to the water to be heated. The heating cycle consists of an evaporator, compressor, expansion valve, condenser and connecting pipework containing a refrigerant.

The evaporator is surrounded by 'fins' to increase the surface with which to draw in heat. The expansion valve releases low-pressure, low-temperature refrigerant liquid into the evaporator, where the lower pressure and resultant lower temperature turn it into a gas. To do this it needs energy to change its state. This energy is drawn from the evaporator and in turn the surrounding air. As the boiling point of the refrigerant liquid is at sub-zero temperatures, the surrounding air always has more energy than the refrigerant. This energy in turn is given up to heat the water as it changes its state back to a liquid in the condenser.

The compressor pressurises the low-pressure refrigerant gas, in turn increasing its temperature before it enters the condenser. The condenser is either a heat-transfer unit, or jacket or piping coiled around the storage cylinder where the refrigerant gas gives up its heat energy to the water to be heated. This high-pressure, high-temperature liquid then enters the expansion valve where it is released as a low pressure liquid into the evaporator to be converted into a gas,

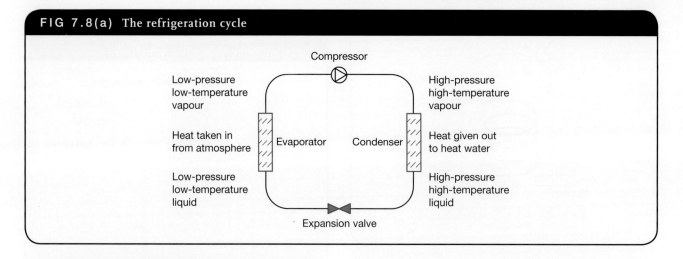

FIG 7.8(a) The refrigeration cycle

Compressor

Low-pressure low-temperature vapour

High-pressure high-temperature vapour

Heat taken in from atmosphere   Evaporator   Condenser   Heat given out to heat water

Low-pressure low-temperature liquid

High-pressure high-temperature liquid

Expansion valve

FIG 7.8(b) Heat pump with separate heating unit utilising heat transfer unit

warm air enters

cooler air exits

heated water returns to the tank

cool water enters from the tank

heat is transferred to the water in the heat pump module

FIG 7.8(c) Heat pump with separate heating unit utilising heat jacket/coil

Fan

AIR IN

Evaporator
Accumulator
Expansion valve

Compressor

HOT WATER OUTLET

Electric booster element (optional)

Condenser

Hot water storage tank

Sacrificial anode

COLD WATER INLET

starting the cycle over again. These cycles continue until enough energy has been transferred to heat the water to the temperature required.

Figures 7.8 (b) and (c) show the two methods of transferring heat mentioned above. The energy used to run the compressor is a fraction of the energy required to heat the water by an electrical element. Heat pumps generally have an auxiliary heating method to boost the temperature of the water where ambient air temperatures cannot provide enough energy to satisfy heat-transfer requirements.

The main advantage of the heat pump is that the sun does not have to be shining for it to work, as it draws the heat energy out of the air. The hotter it is, though, the more efficient the process is. The disadvantage is the compressor can emit noise levels that can be a nuisance in certain situations, such as down the side of a house or next to a bedroom. With advances in technology though, compressors are becoming quieter.

SOLAR WATER HEATERS

The method of using the sun's rays to heat water is rapidly becoming accepted in Australia as a way of providing a domestic hot water supply at an economic cost. This method has been developed by government bodies over recent years and has resulted in the production of a range of simple, straightforward solar water heaters. With the rising cost of energy and the need to be sustainable, solar water heating is becoming more and more popular. It also attracts valuable energy efficiency points required for programs such as BASIX for building applications.

The initial cost of a solar heating system is higher than a normal system, but over a period of years this is recouped by the saving in maintenance and operation costs. The results of tests carried out on a year-round basis of solar water heaters in various areas of Australia show that a supply of hot water adequate for the needs of an average family can be provided by a solar hot water system. These systems also reduce greenhouse gas emissions by using renewable energy.

Solar collectors

The solar hot water system consists generally of one or more solar collectors (Figure 7.9), sometimes called 'absorbers' or 'panels', connected by copper pipes to an insulated storage tank.

When the sun's rays (solar radiation) are received by the absorber, the water contained in the collector is heated and circulated in the tank. This makes the absorber the main component in the solar hot water system. A simple collector consists of a copper plate to which a number of copper tubes are thermally bonded. Insulation is placed behind the copper plate to prevent heat loss; the copper plate is covered by glass which is contained in a sheetmetal or other suitable case.

The absorber, therefore, is a simple heat exchanger that absorbs radiant heat from the sun and transfers this heat to the water within the tubes. Some systems use glycol instead of water which is heated in the collector and passed through tubes in or around the storage tank to transfer the energy collected in the panel to the stored water. The glycol system is completely sealed from the water stored in the storage cylinder.

During normal operation, the collector's temperature can be as high as 37° C above its surroundings and, consequently, heat losses occur. This problem is overcome by the use of adequate insulation to minimise the loss of heat. A minimum of 50 mm of mineral wool or equivalent insulation is placed at the rear of the collector plate and a glass cover is fitted to the front of the unit; the glass cover reduces the loss of heat by convection. The absorber plate

surface is chemically treated to produce a surface having a high absorbance and low emittance in order to enable the plate to absorb the maximum possible heat.

Copper is the preferred material for both the plate and tubes. The use of copper enables the bonding of the plate and tubes to be made without fear of electrolytic corrosion, and because of its high conduction properties, a thin plate can be used. Collectors constructed from copper have an excellent resistance to corrosion, which increases the life of the absorber and renders it virtually maintenance free.

Evacuated tube solar collectors

An alternative to solar collector panels are evacuated tubes. They consist of two high-strength glass tubes. The outer tube is clear and the inner tube has a heat-absorbing coating that draws the heat into the inner tube, and a heat-reflecting coating on the inside to stop the trapped heat from escaping. There is a vacuum between the two tubes which provides excellent thermal insulation (similar to a vacuum flask). A heat probe is placed in the inner tube and extends upwards into a header pipe (Figure 7.10), where the heat is transferred to the water. When heated, the water is circulated (pumped) from the header to the storage cylinder.

The storage tank

On mains pressure solar systems the tank may be positioned above the solar panels or at ground level (Figures 7.11 (a) and (b)). Where the storage tank is located above the collector, the circulation of the water between the tank and collector is maintained by the natural convection currents that are set up when the water in the collector is heated. This natural circulation is known as the thermosiphon principle. Where the cylinder is located at ground level, circulation is provided by a

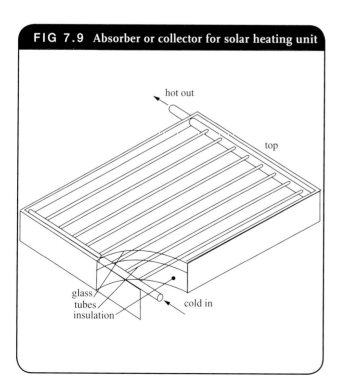

FIG 7.9 Absorber or collector for solar heating unit

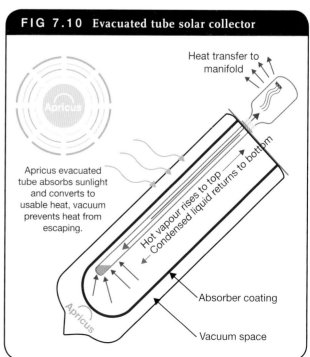

FIG 7.10 Evacuated tube solar collector

circulator (small pump) and controlled by sensors in both the connection to the storage heater and the collector. When a difference in temperature is reached between the heater and the collector, the sensors switch the circulator on and circulate the heated water between the collector and the storage cylinder. An automatic air release valve is installed (Figure. 7.11 (c)) at the collectors where the cylinder is ground-mounted and releases any air that may get caught in the system.

A solar heater should not be expected to heat water without assistance/boosting to required temperatures under all conditions, although in most conditions this is achievable. To account for continuous cloudy weather, a form of boosting device such as an electric element or gas booster needs to be fitted to the storage tank. However, where supplementary heating is necessary to maintain a continuous supply, solar energy can provide

the major part of the heat. All components of solar heating systems are now available commercially in most areas of Australia. Solar collectors may also be coupled to low-pressure ceiling model heaters, as shown in Figure 7.11 (d), and may also be supplied with energy from a slow combustion wood heater in the winter months.

FIG 7.11(a) Hi-line solar heater (glycol model)

FIG. 7.11(b) Low line solar heater with gas booster

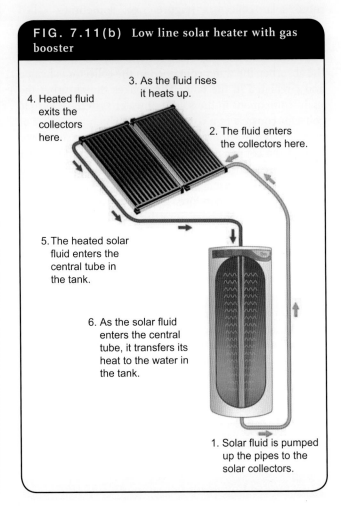

3. As the fluid rises it heats up.

4. Heated fluid exits the collectors here.

2. The fluid enters the collectors here.

5. The heated solar fluid enters the central tube in the tank.

6. As the solar fluid enters the central tube, it transfers its heat to the water in the tank.

1. Solar fluid is pumped up the pipes to the solar collectors.

FIG 7.11(c) Automatic air release valve

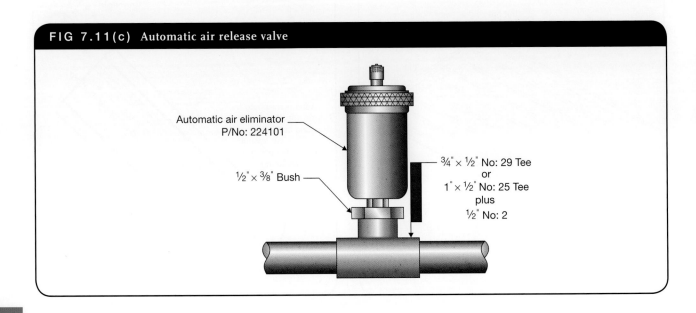

Automatic air eliminator
P/No: 224101

½" × ⅜" Bush

¾" × ½" No: 29 Tee
or
1" × ½" No: 25 Tee
plus
½" No: 2

FIG. 7.11(d) Schematic diagram of a solar water-heating system

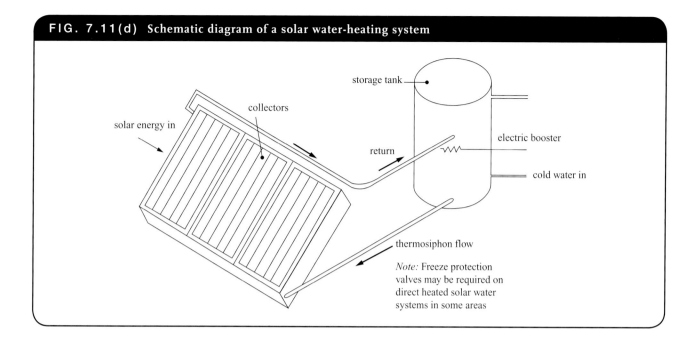

storage tank

collectors

solar energy in

electric booster

return

cold water in

thermosiphon flow

Note: Freeze protection valves may be required on direct heated solar water systems in some areas

BOILERS

Many different boilers are used for the heating of water for domestic purposes. In small households it is quite common for a boiler, in the form of a water jacket or pipe coil, to be incorporated into the firebox of a slow combustion wood heater (Figure 7.12 (a)), cooking range or stove. The boilers are manufactured in a variety of designs and sizes and can be used with many different types of heating media (usually wood). Most of the ranges or stoves are designed specifically for use in the kitchen and are quite pleasing in appearance.

With larger installations, an independent type of boiler (Figure 7.12 (b)) has been designed for the sole purpose of heating water. A cylinder or storage vessel can be used in conjunction with the boiler. The purpose of the cylinder or storage vessel is merely to hold a certain amount of hot water in readiness for instant use.

The cylinder can be installed either vertically or horizontally. This is usually dependent on the size of the cylinder and the space available. If the cylinder is to be fixed horizontally it is good practice to give a slight rise at the end of the connection to the secondary flow to allow for the escape of air from the system.

Capacity of storage cylinder and boiler

The appropriate storage capacity of the hot water cylinder and the heating capacity of the boiler will depend upon the demand for hot water that is likely to be made on the system. Demand varies greatly according to the type of building, the occupants, and the purpose for which the water is required. These factors make it difficult to give precise figures. However, for small domestic installations where the demand for hot water is variable and intermittent, it is normal practice to install a 270 to 315 litre cylinder. Experience has shown that a cylinder of this capacity is sufficient to supply the average household with one bathroom and normal domestic outlets.

Warm water systems

There are two types of systems used to provide warm water. The first system circulates hot water through the flow and return piping, with branches delivering warm water at point of use by means of a temperature controlling device. The branches usually serve one area only and are kept to a minimum length. The circulating hot water prevents Legionella build-up in the flow and return piping; the minimum branch length also assists with this, although the water needs to be periodically tested for any build-up.

The second system circulates warm water through the flow and return piping. Warm water is the perfect breeding ground for Legionella and other bacteria and viruses, so it needs to be disinfected by ultra-violet light (UV) (Figure 7.13(a)) or other means to control any build-up. The water heaters are set at normal temperatures and the water feeding the flow piping is controlled by large thermostatic mixing valves (Figure 7.13(b)). Under flow conditions the valves mix the hot and cold together to provide warm water. Under no flow conditions the warm water is circulated through the warm water piping and partly through the hot water system, which replenishes the energy lost in the flow and return piping.

Refer to the relevant standards for details of warm water systems, including testing of water and maintenance requirements.

Multiple installations (equaflow)

Where large volumes of hot water are required using a centralised system, water heaters may be 'manifolded' together to act as one unit. To ensure all heaters are utilised

FIG 7.12(a) Solid-fuel water heating system (can be used with a slow combustion wood heater or stove)

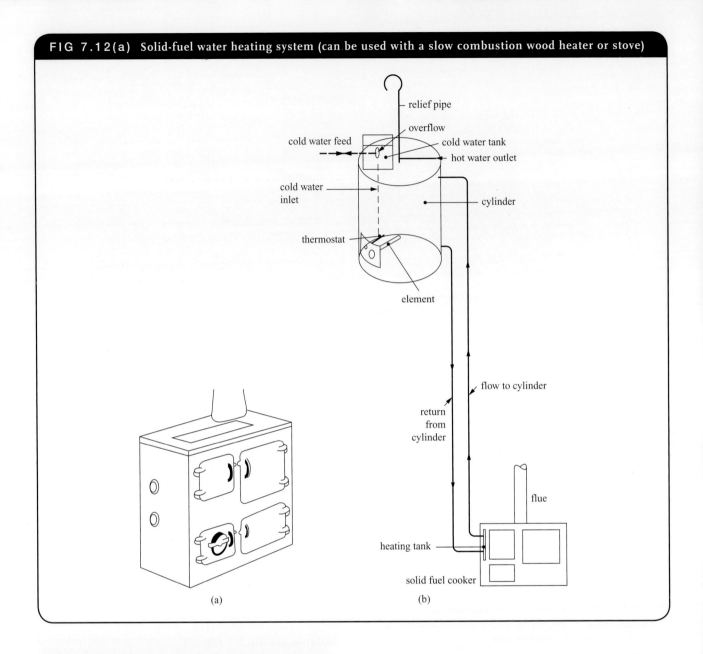

(a)

(b)

relief pipe
overflow
cold water feed
cold water tank
hot water outlet
cold water inlet
cylinder
thermostat
element
flow to cylinder
return from cylinder
flue
heating tank
solid fuel cooker

FIG 7.12(b) Raypak gas-fired boiler

equally, the piping layout must ensure that the first heater to receive the cold water supply is the last one to deliver the hot water (equaflow). This prevents a short-circuit (i.e. one heater doing all the work). Where the hot and cold water enters from the same side of the installation, either the cold or hot water piping must 'reverse return' so the distance travelled through each heater is the same (Figure 7.14). To ensure that equaflow principles have been followed the same distance of flow through each heater must be exactly the same distance from where the cold water enters the cold water manifold and the hot water exits the hot water manifold.

Either storage (Figure 7.14), continuous flow heaters (Figure 7.15), solar heaters (Figure 7.16) or heat pumps (Figure 7.17) may be manifolded together to provide large volumes of hot water. Manifolded continuous flow heaters usually come preassembled from the factory.

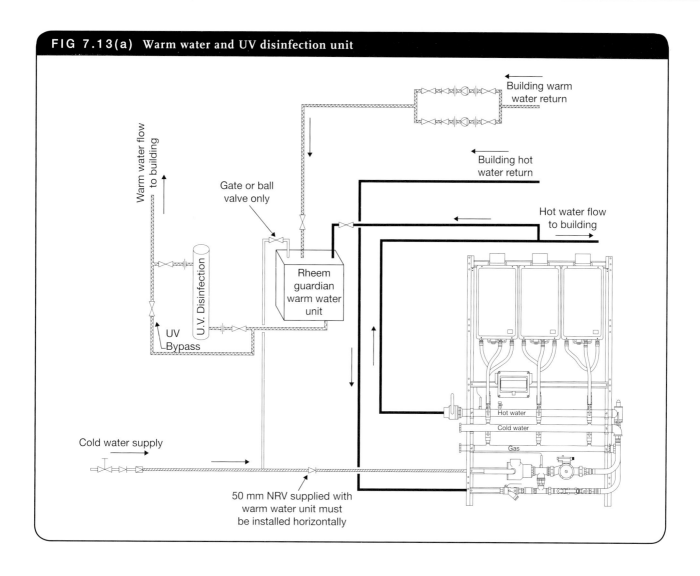

FIG 7.13(a) Warm water and UV disinfection unit

Building warm water return

Warm water flow to building

Gate or ball valve only

Building hot water return

Hot water flow to building

U.V. Disinfection

Rheem guardian warm water unit

UV Bypass

Hot water

Cold water

Gas

Cold water supply

50 mm NRV supplied with warm water unit must be installed horizontally

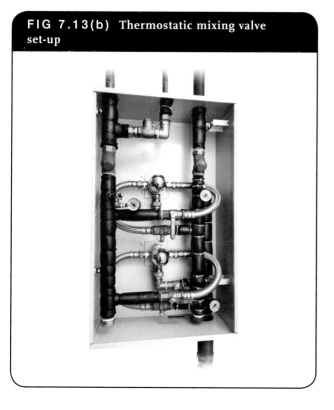

FIG 7.13(b) Thermostatic mixing valve set-up

CONSTRUCTION MATERIALS FOR HOT WATER SYSTEMS

Copper

Copper has been used in the construction of hot water tanks since the evolution of storage units. It is used for low-pressure ceiling model storage systems and cistern-fed floor model heaters, and the cold-feed supply cistern for both. Many systems that utilise wetbacks and/or solar panels are connected to copper cylinders (Figures 7.11(d) and 7.12(a)).

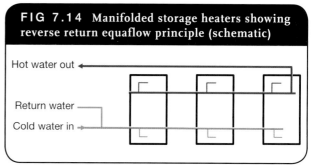

FIG 7.14 Manifolded storage heaters showing reverse return equaflow principle (schematic)

Hot water out

Return water

Cold water in

FIG 7.15 Continuous flow manifold

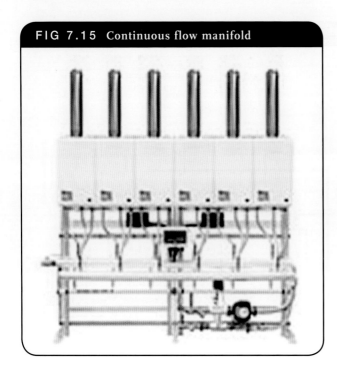

Silicon bronze (Cusilman)

Silicon bronze (Cusilman) is an alloy of copper, with copper's corrosion resistance and the strength of mild steel. It was once regarded as the best material for mains pressure units (although it would not take full mains pressure) but is no longer used for water heaters. A great many units were made with this material. Properly constructed and of the correct thickness, they provided many years of use and some may even be still operating today.

Copper-lined steel cylinders

Copper-lined steel cylinders were manufactured by a number of Australian companies. They were a combination used in heaters to obtain strength and corrosion resistance in mains pressure units. A copper lining was inserted into a thickness of mild steel. Copper-lined cylinders are no longer manufactured and have been replaced by glass-lined cylinders (vitreous enamel lining).

FIG 7.16 Commercial solar installation

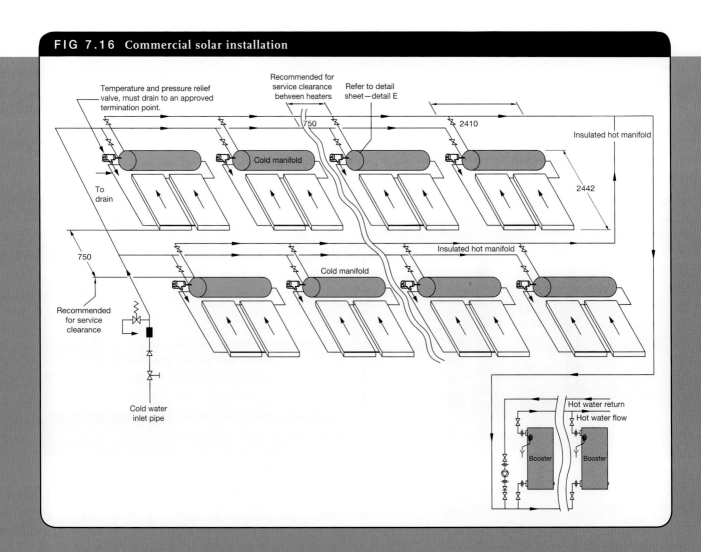

FIG 7.17 Heat pump commercial installation for both hot water and warmwater

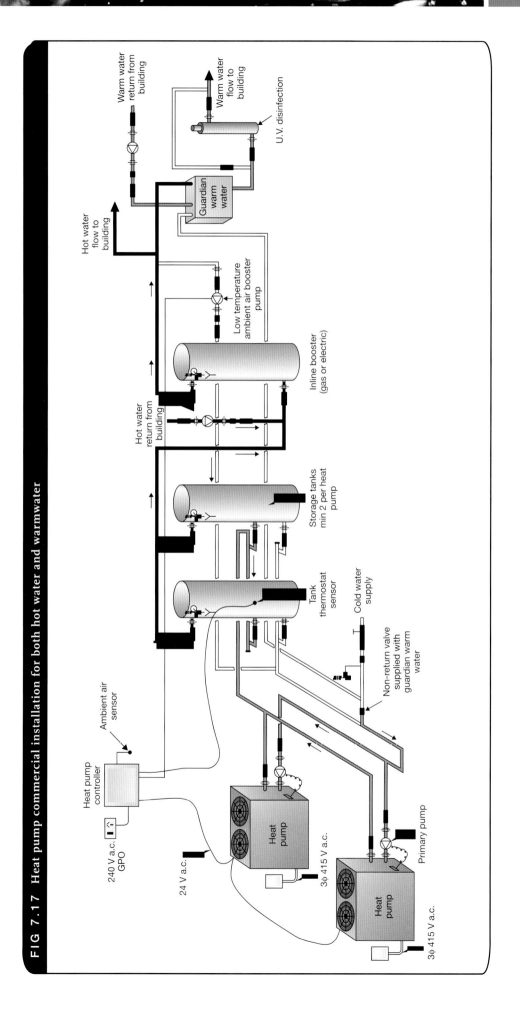

Glass-lined steel cylinders

Glass-lined steel cylinders are also manufactured by a number of Australian companies. These mains pressure units were originally offered as a cheap alternative to copper-lined steel units or silicon bronze construction but are now the mainstay of mains pressure water heaters. The lining on the glass cylinders is a bonded vitreous enamel, and is very similar to the coating used on cookware and metal baths.

The glass lining is protected against flaking and the subsequent rusting of bare steel by soluble protective anodes inserted in the tanks. Care must be taken that the anode is suited to the water quality it is exposed to and is replaced when necessary (recommended to be replaced every five years). Glass linings are also soluble in some waters and under high-temperature conditions.

The life of a cylinder can be extended by the addition of a second glass lining that can withstand higher temperatures, such as produced by water heaters serving commercial kitchens. This high-temperature enamel is bonded to the first enamel coating, as it won't successfully bond to the bare steel—hence the need for a double lining. This second lining usually has the added benefit of extending the warranty of the heater.

Stainless steel

Stainless steel is used for mains pressure water heater cylinders to provide strength and long service life. Several solar heating companies use stainless steel as their preferred option. Stainless steel is free from any coating and does not require a sacrificial anode. Marine grade 316 stainless steel is used for hot water cylinders (Figure 7.18) by most manufacturers and has a natural resistance to rust and corrosion. It is often backed by a 10-year warranty as a testimony to its longevity.

TESTING AND COMMISSIONING OF HOT AND COLD WATER SYSTEMS

As soon as possible after installation, all systems should be thoroughly flushed out to remove any foreign matter such as flux residue, jointing compound and dirt. After flushing, the completed system should be tested and inspected by the plumber under working conditions of pressure and flow, and when all draw-off points are closed

FIG 7.18 Stainless steel cylinder

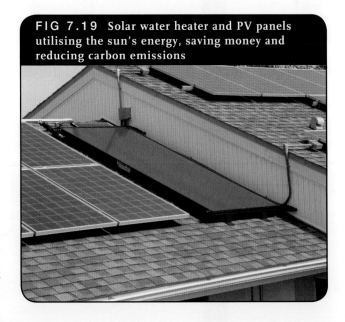

FIG 7.19 Solar water heater and PV panels utilising the sun's energy, saving money and reducing carbon emissions

SUSTAINABLE PLUMBING: RENEWABLE SOURCES OF ENERGY

The need to be aware of our carbon footprint is becoming increasingly important. Energy from renewable sources such as heat pumps and solar energy are becoming more a necessity than a luxury. The use of energy-intensive forms of heating water will become a thing of the past. The greater the demand for these alternative systems the more attractive the cost will be. Sights such as Figure 7.19 will become the 'norm' as we move towards saving and providing energy from alternative sources such as the sun—which generates energy for free!

the system should be absolutely watertight. Each draw-off point should be opened and tested for the rate of flow.

Pressure testing

According to AS/NZS 3500 a service, hot and cold, is to be tested to 1500 kPa, for a period of not less than 30 minutes. Before hydrostatic testing is employed, the piping system is to be cleaned and flushed to remove all the foreign matter.

Where a pressure test is employed it may be necessary to isolate items of equipment from the test, such as the boiler and hot water cylinder relief valves and other ancillary equipment not capable of withstanding the pressure test. Where these items are removed, blanking flanges or plugs must be used, or a make-up piece of pipework installed.

Where thermal insulation is used, the test should be made before the insulation work is complete and while all joints are exposed. It may also be necessary to carry out testing on sections of pipework before completion of the whole installation where these are fitted in ducts, chases in brickwork and trenches and are concealed from view.

Methods of testing

The most common method of testing is the hydrostatic test, in which the pipework is water-tested under pressure. This can be achieved in two ways:

1　Direct pressure, where the water pressure in the water authority's main is used to pressurise the system; this type of test is normally used on cottages or extensions to existing services. Pressure is boosted by a test bucket.

2　Indirect pressure, where because of construction or installation problems such as back-filling of trenches or cement rendering of chases, portions of pipework would have to be tested before completion of the installation. The test involves filling the pipework and a test bucket with water; operating a hand pump on the bucket to bring the pressure up to the required reading on the attached pressure gauge; turning off the control valve on the bucket and observing the pressure gauge for any drop in pressure. After the test has been applied for the required time, the control valve is turned on allowing the pressure to be released.

NOTE: An alternative would be to pressurise the system using compressed air. This method saves water and minimises damage due to faulty joints; however, leaks are harder to identify if the pipework is hidden.

Commissioning

After the installation has been completed, tested and found to be satisfactory, the job of commissioning remains. In most installations, this consists of the procedure described in the following box.

Test pumps are ideal for the pressure testing of water lines and containers up to 60 bar with very high testing

INSTALLATION PROCEDURE

1.　Remove protective plastic coating on chrome piping, plastic handles, etc.

2.　Remove labels and protective coatings from plastic-coated items, such as stainless steel sinks.

3.　Fill the system with water before applying the heating medium to the hot water service.

4.　Fully purge the air from the system.

5.　Test the operation of valves, cisterns and pressure-relief valves.

6.　Check the stored hot water and the delivery temperature of the hot water in accordance with AS/NZS 3500.4.

7.　Check the water level in a gravity-style hot water system.

8.　With the inlet isolating valve fully open, check the flow rate at the outlet points.

9.　Check the operation of the circulating pump and the flow and return temperatures.

10. Check for vibration, noise and/or water hammer.

11. Make the operating instructions available to the owner or the occupier of the building.

standards. The RP50-S features a drag indicator that displays the preset pressure so that a loss of pressure in the system can be detected more easily (Figure 7.20). Delivering reliable and precise testing results, the dual-valve system enables self-testing of the pump, as well as a precise fine adjustment of the pressure. The testing hose is comprised of steel mesh, which prevents reading mistakes that can happen when the hose expands during operation. The 12-litre testing pump has a weather and cold-resistant, steel-coated container.

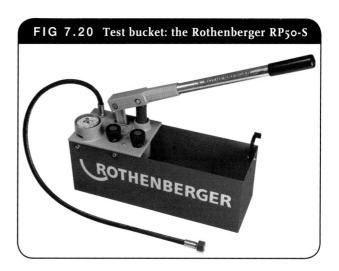

FIG 7.20　Test bucket: the Rothenberger RP50-S

FOR STUDENT RESEARCH

Visit your local plumbing hardware store. Enquire about, and view, all the equipment needed to test and commission hot and cold water systems.

- AS/NZS 3500 Part 4: Heated water services provides acceptable solutions to installing water heating systems that comply with the Plumbing Code of Australia and the New Zealand Building Code. AS/NZS 3500 cites many other standards relevant to the materials and their installation. Refer to these standards to further your understanding of heated water services and hot water piping systems.

Visit the following manufacturers' websites for more information on water heating systems

- www.apricus.com.au
- www.aquamax.com.au
- www.chromagen.com.au
- www.dux.com.au
- www.edson.com.au
- www.edwards.com.au
- www.endless-solar.com.au
- www.everlastwaterheaters.com
- www.geothermalaustralia.com.au
- www.hillssolar.com.au
- www.pivotstove.com.au
- www.quantumenergy.com.au

- www.rheem.com.au
- www.rinnai.com.au
- www.sanden-hot water.com.au
- www.solahart.com.au
- www.solarsavers.com.au
- www.stiebel.com.au
- www.wilsonhotwater.com.au

The following websites explore energy-saving ideas and low-energy water-heating alternatives:

- www.ata.org.au/sustainability/solar-hot water
- www.climatechange.gov.au
- www.ecosmart.com.au
- www.energyrating.gov.au/products-themes
- www.resourcesmart.vic.gov.au/for_households_3096.html
- www.saveenergy.vic.gov.au/atwork.aspx
- www.tasgas.com.au
- www.thenaturalchoice.com.au

Australian Standards

AS/NZS 3500.1: 2003 Plumbing and drainage—Water services

AS/NZS 3500.4: 2003 Plumbing and drainage—Heated water services

ON-SITE STORIES 7.1

TRANSFERABLE SKILLS

Steve Eckert, *Job title: Director of Eco Building Supplies, South Australia*

There has been a significant advancement in the hot and cold plumbing market in the last 20 years. The plumbing industry has advanced from using galvanised steel pipe to copper and now polymer/composite pipe, which currently has over 15 different types of systems in Australia alone.

In 1996, I was working on the then new Mount Gambier Hospital. The plumbing company I was working for was the first commercial company in South Australia to use a hot and cold polymer system called REHAU. There was only one other polymer system being used at the time in South Australia, but it was not what I deemed a failsafe product. I asked the REHAU

sales representative who was visiting on site if they had any designated representatives in South Australia. He told me the market was too small. A few months later I called the Victorian state manager to arrange a meeting and 18 months later I started the South Australian REHAU office in Adelaide with support from the Victorian branch. The South Australian branch grew and I remained with the company for over 10 years. It is still a market leader today.

The moral to this story is that completing your plumbing apprenticeship provides a strong foundation and can set you up for life. The plumbing industry offers significant diversity and can take you anywhere you want it to. I have been fortunate to work in Canada, the United states and Papua New Guinea and have run my own business with a business partner for the last four years.

Knowing that I have a trade behind me is a safety net for the future. The knowledge gained during the training has been invaluable and is transferrable to work anywhere in Australia or overseas. You are only ever limited by your own imagination. Good luck!

Rural water supply

INTRODUCTION

People who live in areas serviced by a reticulated water supply system often take for granted the convenience offered by this system and the amount of potable (drinking) water used by the average family. This is not the case, however, for people who live in small rural communities or on isolated farms who do not have a reticulated supply provided by a water supply authority. These people are frequently faced with shortages of drinking water and, as a result, actively engage in water conservation from a very early age.

Australia is generally accepted as being the driest continent on earth and elaborate steps are taken to ensure that an adequate supply of water is available throughout the year so that normal standards of water usage can be maintained. Despite careful calculations on user requirements and the provision of storage facilities, people in arid areas often find it necessary to purchase water from other sources to supplement their own storage systems, especially during the prolonged periods of drought that occur at irregular intervals.

The purchase and transportation of water is an expensive and time-consuming exercise. For this reason, extreme care should be used when calculating the requirements and storage capacity for a rural installation.

WATER QUALITY

It is essential that the quality of water at the source of supply is suitable for the purpose for which it is intended. It is important to have all water tested by qualified chemists who will be able to certify its suitability.

Water collected from precipitation is usually pure enough for human consumption without further treatment. Groundwater, however, should always be regarded as suspect until tests have been carried out. This is especially true of groundwater from a new or previously untapped source.

Analysis of water intended for watering stock or irrigation of crops is usually carried out by the various state departments of agriculture. In some areas the local or shire councils are also equipped to carry out these tests. This is not the case, however, with water intended for human consumption. The analysis and testing of potable water usually comes under the jurisdiction of state health departments.

The testing authorities generally require two samples from each source and each sample should contain at least 500 mL. Containers used for samples should be thoroughly washed and fitted with watertight screw caps.

SOURCES OF SUPPLY

Traditionally, the majority of rural installations obtain their potable water by collecting and storing the precipitation that falls on the roofs of farm buildings. When roofs are to be used for this purpose, it is imperative that they be kept clean and as free as possible from substances that may cause contamination of the water supply.

In rural areas, the most commonly used roof covering materials are zinc-coated steel, aluminium, Colorbond and, to a lesser extent, terracotta or cement tiles. The water is collected from the roof in gutters and conveyed via downpipes to the storage tank, which may be situated either above or below ground.

Maintenance of rainwater tanks for acceptable water quality

The quality of the water stored in a rainwater tank depends on correct installation and design, incorporating a sensible maintenance program of the tank and the catchment area (Figure 8.1). The tank should be covered at all entry points with mesh screens to prevent leaves, vermin and mosquitoes from entering it. The fitting of a first flush device is very important; the device prevents contaminants from entering the tank by discarding the first portion of rainfall, which will also protect rainwater pumps and household appliances. Figure 2.13(a) in Part 2 illustrates the above points.

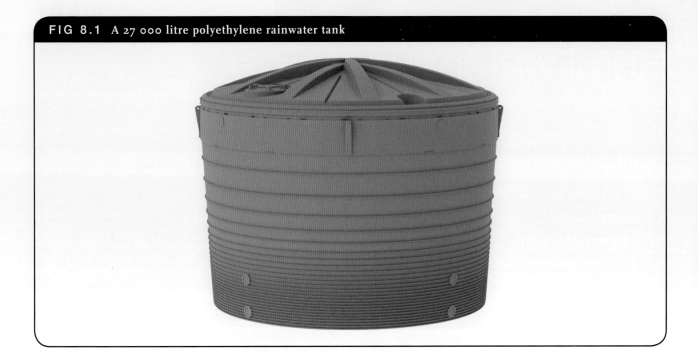

FIG 8.1 A 27 000 litre polyethylene rainwater tank

Roof materials for water harvesting

The roof surface is critical to ensure the collection of good quality rainwater, as some materials— such as lead or bitumen-based paints—are harmful for use with drinking water. Check local council requirements, and information from manufacturers, to determine if the material is suitable for rainwater collection.

The most commonly used roofing materials suitable for potable water are:

- Colorbond
- galvanized steel
- Zincalume
- glazed tiles.

If the water being collected is not being used for drinking, the material being used is irrelevant—the water collected would be safe for gardening, car washing and other uses around the home.

Development of groundwater supplies

Water trapped beneath the earth's surface, known as groundwater, is an important source of potable water for rural communities. Groundwater is not uniformly distributed throughout Australia and there are considerable variations in quality, depth and yield, even over short distances. Groundwater quality can vary from almost as pure as rainwater to water that contains a higher salt concentration than the sea. The depth at which groundwater may be encountered varies from 2 metres to 1500 metres, and yields that range between practically zero to over one million litres per hour are not uncommon.

These variations are due to many factors including climate, topography and the most dominant factor, geology. Geology determines the storage condition of the groundwater, which in turn controls the rate at which

water can be extracted from the aquifer and the quality of the water contained in the aquifer.

Types of groundwater

There are two basic types of groundwater: confined and unconfined. The disposition of the aquifer is the factor controlling the classification. If the water contained in the aquifer has a free surface, the water is classified as unconfined. This means that the water is subjected to atmospheric pressure and if a bore or well is sunk to the water table the water will not rise up the bore casing above the water table in the aquifer (Figure 8.2(a)).

In situations where the aquifer is sandwiched beneath an overlying or confining formation, the water from this type of structure is classified as confined groundwater and is subjected to pressures exceeding atmospheric pressure. If a bore or well is sunk to the level of the aquifer in this situation, the pressure will force the water from the bore to a height equal to the pressure in the aquifer. The resulting bore will be termed 'artesian' (Figure 8.2(b)).

Springs

These occur generally at or near the base of a natural slope and are created by water percolating through the subsoil, down the slope. At the base of the slope the subsoil becomes saturated and water emerges at the surface. The yields from this type of spring are generally unreliable and will vary considerably with seasonal changes.

Springs created when the aquifer is fractured or faulted causing water to seep from the surface are usually a much more reliable source of water as they are not subject to the same seasonal variations in conditions.

Wells

These are excavated, either by hand or by mechanical means, to enable groundwater to be extracted. They are

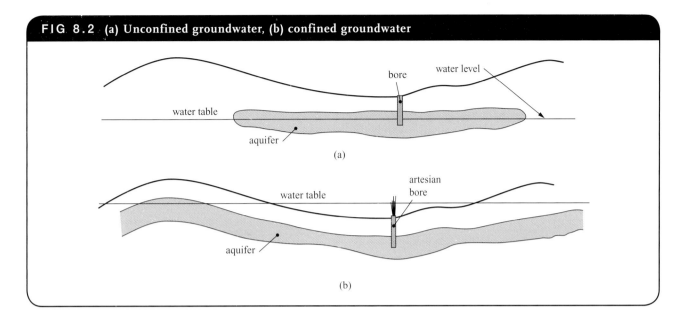

FIG 8.2 **(a) Unconfined groundwater, (b) confined groundwater**

used where the water-bearing strata are relatively shallow and are usually lined with concrete liners to prevent the collapse of the surrounding ground or to prevent the ingress of surface water, which is more likely to be contaminated. As an added safeguard the well liner should be left protruding above the surrounding ground surface so that surface run-off will not enter the well during heavy rains. Alternatively, the well may be completely capped.

The section of the well liner passing through the water-bearing strata is usually constructed of hardwood runners spaced to allow the water to enter the well. These runners are held in position with steel bands, to which the runners are bolted (Figure 8.3).

Narrow encased well

A water well can also be constructed by mechanically boring a narrow vertical shaft into the ground, as illustrated in Figure 8.4. The cylindrical casings can be constructed of materials such as PVC-U pipe with preformed metal cylinders. The vertical pipe casing is used to keep the bore hole from caving in, and the surface contaminants from entering the bore hole, and protects the installed submersible pump from drawing in sand. Over-pumping of the well at times of peak demand can cause sand intrusion, which causes a clogging of the system and wear to the pump.

STORAGE REQUIREMENTS

Rainfall

In the majority of rural installations in Australia, water for potable purposes is collected from the roofs of buildings or from artificial catchments. This method of obtaining water is totally dependent on the rainfall of the area to supply all the potable water requirements for the installation in question. Generally, a roof or artificial catchment, if properly constructed and maintained, is capable of collecting 80 per cent of the rainfall that falls on it. If these catchments are not carefully maintained, the 80 per cent collecting potential can be reduced *dramatically*.

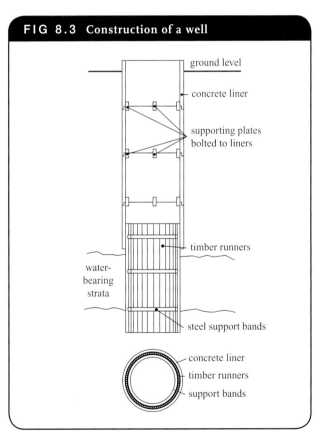

FIG 8.3 **Construction of a well**

Type and purpose of storage

Type and purpose of storage are interdependent factors; the ultimate use of the water usually dictates the type of storage used. This text is primarily concerned with potable drinking water storage. If, however, the requirements of an installation included crop irrigation and stock watering, these uses would dictate the type of storage—generally ground tanks where seepage and evaporation are likely to occur.

FIG 8.4   A typical encased well with a submersible pump

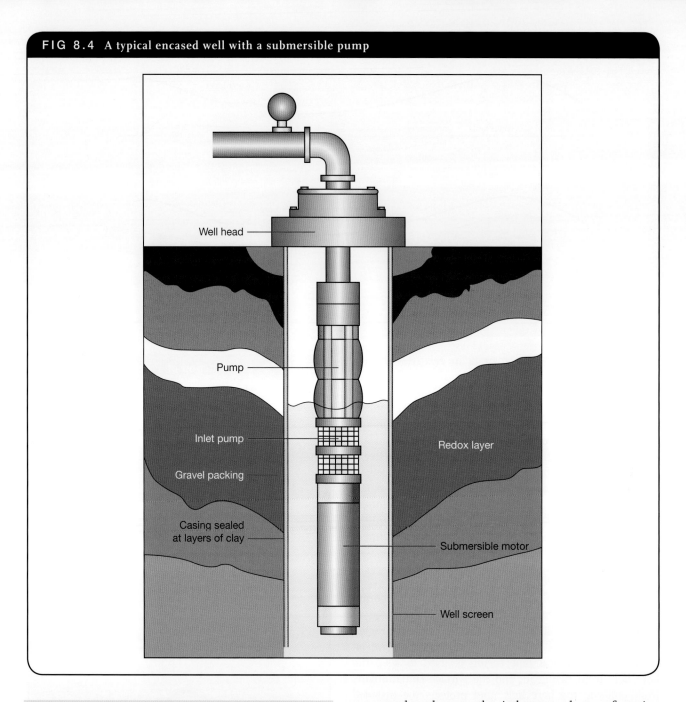

Well head

Pump

Inlet pump

Gravel packing

Casing sealed
at layers of clay

Redox layer

Submersible motor

Well screen

DETERMINING STORAGE REQUIREMENTS

When determining the storage requirements for a rural installation, several factors must be considered:

1. the average rainfall of the area
2. the duration of storage without replenishment
3. the type of storage to be used
4. the purpose for which the water is to be used
5. the individual requirements of the installation.

The rural sector

The amount of water used in the agricultural sector is far greater than that used in urban areas. The agriculture sector consumes 65 per cent of the bulk net water consumed each year, the industry and manufacturing sectors 23 per cent and the household sector about 11 per cent. In addition, water consumption by irrigated agriculture has been steadily increasing over the last two decades. Groundwater provides domestic water for more than one million Australians in the majority of rural communities and 60 per cent of the continent relies on it for all uses except drinking water.

The urban sector

Domestic consumption has increased significantly, as a consequence of increasing population and rising per capita use. However, it varies significantly from place to place, according to rainfall, number of rain days, mean temperatures and humidity, availability of water, pricing and education. Annual household consumption of water in Darwin is about 2.5 times that in Sydney.

There have been significant reductions in water consumption over the past decade, due to educational programs in conservation. But the overall demand is gradually increasing due largely to increasing population. Many cities and towns in Australia will need to find new sources of water unless there are significant improvements in water conservation.

For example, careful design of the pipework layout in dwellings will enable water from several sources to be interconnected so that the purest water required for potable purposes will not be wasted on washing and flushing sanitary appliances.

STORAGE TECHNIQUES

Water for rural installations may be stored in numerous ways depending on its ultimate use. Water intended for stock or irrigation is usually stored in earth tanks or ponds at ground level, while water for human consumption is generally stored in tanks manufactured from steel or concrete situated either above or below ground.

Tank materials

Tanks are available in a variety of materials.

Plastic tanks

Plastic polyethylene tanks are durable, UV resistant, made from premium food-grade material, light and easy to transport, available in many colours and suitable for both above- and below-ground installations (See Table 8.1). Rural and industrial tanks come in a range of sizes from 55 to 50 500 litres.

TABLE 8.1 Comparison of above-ground and in-ground plastic tanks		
Type	**Features**	**Considerations**
Above- ground	• Lightweight and easily transported— good for smaller tanks. • Flexibility in shapes and colours. • Easy to install, service and maintain.	• Despite having UV inhibitors, best placed in shade.
In- ground	• Good choice of materials and clever design maximises strength, minimises depth and increases practicality. • Anti-hydrostatic lift measures, such as good design features, anchoring or ballast will be needed as pressure from high groundwater can force it out of the ground. • Need to protect tank water from overflow surges running back into the tank.	• Load-bearing can be limited. • Need to be integrated into system that can include driveways, etc.

Metal tanks

Metal tanks are relatively easy to transport, can be custom made, and are corrugated or straight rolled. Metal tanks can be made from the following materials:

- galvanised steel
- Aquaplate or Colorbond
- Zincalume
- copper and stainless steel.

Table 8.2 compares the features of some different types of metal tank.

SUSTAINABLE PLUMBING: EMBODIED ENERGY AND GREENHOUSE GAS IMPACT

To assess the environmental impact of rainwater tanks, the *total embodied energy and emissions* involved in their manufacture should be taken into account. This allows the comparison of the environmental performance of different rainwater tank materials such as plastic, concrete and metal.

The assessment would include energy and emissions generated by extraction of raw materials and transportation of materials to the production site (e.g. to a steel works or cement producer), as well as the actual production of the materials. The assessment would also consider the transportation of the tank to the customer's site, the installation, and the disposal of the tank at the end of its operational life. Ongoing energy use of the tanks pumps and treatment processes also need to be considered.

TABLE 8.2 Metal tanks

Type	Features	Considerations
Galvanised steel	• The classic outback tank. Readily available and relatively easily transported. • Galvanised steel performance can be improved with rust-resistant coatings such as Zincalume or Aquaplate. • Easy to install, service and maintain.	• Initial corrosion of galvanised steel normally creates a thin adherent film that coats the interior surface of the tank and provides protection against further corrosion. Cleaning should not disturb this film. • Avoid copper or copper alloy fittings (brass and bronze) connected directly to steel tanks as this causes corrosion.
Aquaplate/Colorbond	• Aquaplate steel has a polymer skin bonded to a corrosion-resistant galvanised steel base. • Colours can match roofs and fences. • Easy to install, service and maintain.	• The polymer coating is not resistant to prolonged exposure to sunlight so tanks must have a top cover in place at all times. • Avoid copper or copper alloy fittings (brass and bronze) connected directly to steel tanks as this causes corrosion.
Stainless steel	• High resistance to corrosion, staining and bacteria. Available as a garden design feature and in a range of shapes and sizes. • Easy to install, service and maintain.	

Galvanised steel tanks

The traditional corrugated galvanised steel water tank is a familiar sight throughout Australia and still enjoys immense popularity; although modern manufacturing techniques and improved installation procedures have increased the use of concrete tanks, which are either precast or cast in situ.

There are two basic types of galvanised steel water tank used throughout Australia. They are the riveted and soldered tank, and the bolted tank. The riveted and soldered tank is generally used for small capacities—from 4.5 to 21.5 kilolitres—while the bolted type is suitable for much larger capacities up to 177 kilolitres (Figure 8.5).

Both these types of tanks are available in kit form for ease of transportation. The kits are complete, even to the extent of supplying soldering flux for the soldered type. Smaller capacity tanks may be purchased already assembled.

Galvanised steel tanks are suitable only for installations at ground level or elevated on steel or timber stands. When installed at ground level they should be slightly raised on a mound of earth that is retained inside a special retaining ring of material similar to that of the tank. This mound ring is designed to raise the base of the tank so that moisture from the surrounding ground will not be trapped beneath the tank and accelerate the corrosion of

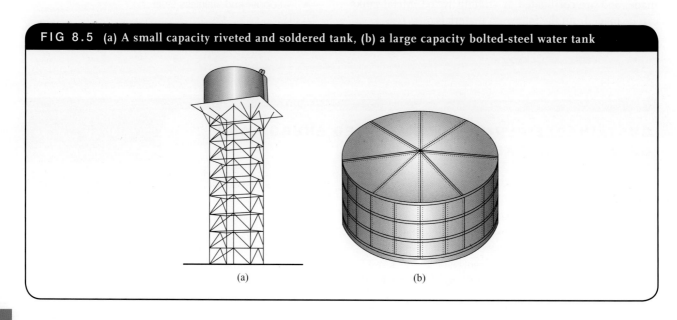

FIG 8.5 (a) A small capacity riveted and soldered tank, (b) a large capacity bolted-steel water tank

(a)   (b)

the tank base. A layer of tarred felt should be placed under tanks which come into contact with sand or soil.

The life of a galvanised steel water tank depends on a number of factors and careful consideration of these factors may extend the life of a tank by many years. Tanks should be installed whenever possible in a shady situation, preferably on the southern side of a building and never directly beneath overhanging trees as falling leaves, insects and bird droppings may contaminate the stored water. Sunlight falling directly on a tank for extended periods may create temperature differentials in the stored water, which tends to encourage internal corrosion.

Water storage tanks should always be fitted with a top to prevent contamination and to screen direct sunlight, which will accelerate the growth of algae. The majority of algae types found in stored drinking water are not detrimental to health and do not reduce the potability of the water. However, some produce acids during their normal process of growth that may be detrimental to the zinc coating on the tank.

All inlet and outlet fittings on galvanised water tanks should be of a material that is compatible with the galvanising. Copper, brass or bare steel should not be placed in contact with the zinc coating unless it is tinned, otherwise the zinc coating will be destroyed adjacent to the fitting. Tank inlets should be installed so that they protrude well into the tank and water entering the tank does not run down the inside and damage the galvanising.

NOTE: Zincalume or copper components will lead to early corrosion of galvanised tanks.

Galvanised steel tanks are used to store water from many sources such as rain, river, bore and dam water. Zinc is not capable of resisting corrosion by itself. However, if the formation of a surface oxide is carefully controlled it will form a very tough adherent film that is capable of reducing further corrosion to a very low level.

This film formation is difficult to obtain with water that is chemically pure, such as rainwater. For this reason, chemical additives may need to be added to the water to assist the formation of protective oxides. A substance commonly used for this purpose is made from soluble zinc metaphosphate and zinc-calcium metaphosphate; it dissolves into the water during the initial filling of the tank and assists in the formation of the protective oxide.

Bore, well, dam and river waters vary considerably in composition. Some may be aggressive to zinc-coated steel, while water from another similar source may be protective. This variation in the character of water is caused by the soil types through which the water passes. Water from previously untried sources should be tested before storage in a galvanised steel tank as special surface protection for the tank may be required.

Concrete tanks

The popularity of concrete tanks has increased over the past years due to improvements in transportation and unloading techniques and the wide variety of mastic jointing compounds for on-site jointing.

There are two common methods of construction used for concrete tanks, 'cast' and 'sprayed'. Precast reinforced concrete tanks are available in capacities up to 23 000 litres. Tanks of this capacity are manufactured in two halves that are joined together on site using a mastic jointing material. Sprayed tanks are usually of a smaller capacity than cast tanks and are manufactured in one piece.

Concrete tanks have the added advantage that they can be installed either above or below ground. They require a level area on which to stand with a minimum of 50 mm depth of bedding material (usually sand) to ensure accuracy in levelling. They are fitted with 50 mm brass BSP threaded bosses to receive inlet, outlet float valve and pressure unit assemblies.

Table 8.3 compares above-ground and in-ground concrete tanks.

TABLE 8.3 Concrete tanks

Type	Features	Considerations
Above-ground	• Good for larger capacity tanks. • Not usually used in urban settings. • Lime from cement softens the water. • Easy to install, service and maintain.	• Heavy, so needs strong foundations. • Tanks can be poured on site. • Needs sealing for maximum water retention.
In-ground	• Inconspicuous large tanks. • Lime from cement softens water. • Sealed with latex or other lining.	• Can be placed in traffic areas as they can be designed as load-bearing structures.

Innovative tanks

Manufacturers have responded to change in a creative way, improving processes and products in style and colour, with innovative solutions for harvesting and storing rainwater without taking up valuable space. Innovative tanks are designed to fit in tight spaces, around obstacles such as columns and posts, within homes and businesses. This reduces the demand on other water supply options, such as reservoirs. Different materials are used to create enclosed storage tanks such as slimline bladder styles, while retaining a sustainable focus for harvesting rainwater. Table 8.4 provides a comparison of different types of tanks.

Tanks in series

Where a multiple tank installation is required, care should be exercised in the method of connection or interconnection of the tanks. For example, it has been found that in multiple tank installations the second or subsequent tanks in the line have a very much shorter life than the first tank. It is therefore important to have a separate inlet and outlet for each tank. The reason for the shorter life of series-connected tanks appears to be a variation in the purity of the water in the various tanks. For this reason the methods of connection shown in Figure 8.6 should be used whenever a multiple tank installation is required.

SUPPLY SYSTEMS

There are two basic supply systems used for rural water supplies: the gravity system and the pressure system.

Gravity system

This type of installation relies on the 'head' or height of the stored water level to provide a satisfactory pressure of water at tap outlets. To achieve this, the storage vessel must be raised on a platform so that the lowest possible water level in the storage tank will provide a satisfactory flow to the taps.

The pressures available from this type of system are very low; for example, a tank installed three metres above a tap will only produce approximately 30 kpa. For this

TABLE 8.4 Innovative tanks

Type	Features	Considerations
Water walls	• Generally plastic or metal, good for limited ground space.	• Can be difficult to clean and maintain in protected areas when in exposed sites.
Bladders	• Innovative under-deck or under- house bladder made from tough materials. • Can collect from a number of drainpipes unobtrusively and utilise previously wasted space.	• Ongoing maintenance and access needs to be considered.
In slab	• Used like a waffle pod, where the concrete slab is sitting on and around a series of boxes (or pods) set out in a grid pattern. • Each in slab is approximately 600 litres.	• Used in new homes or extensions; cannot be retrofitted. • Ongoing maintenance and access needs to be considered.
Fibreglass	• A food-grade coating on the interior surface is cured before the tanks are offered for sale. • Lightweight and strong. • Flexibility in shapes and colours. • Relatively salt resistant, so good in coastal locations. • Relatively easy to repair.	• Despite having UV inhibitors, better to be placed in shade.

FIG 8.6 (a) Double tank installation. The overflow from one tank should not be used to fill others. (b) Each tank should have its own inlet and outlet.

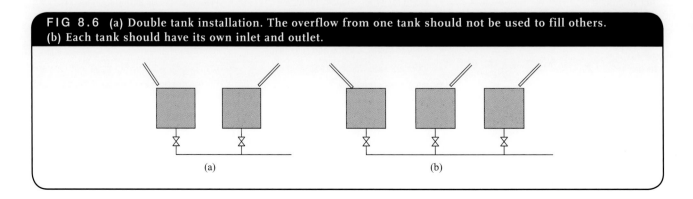

(a)   (b)

FIG 8.7 Assembly of two-part precast concrete tank

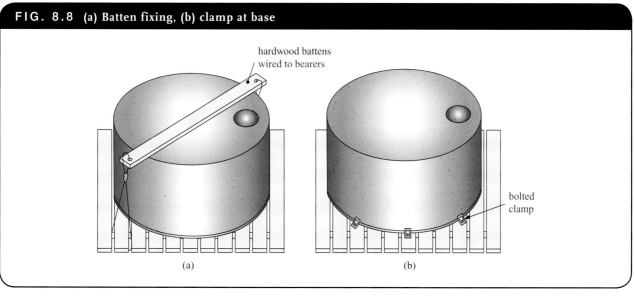

FIG. 8.8 (a) Batten fixing, (b) clamp at base

hardwood battens
wired to bearers

bolted
clamp

(a)

(b)

reason, extreme care should be exercised when sizing pipework and designing pipework layouts. Conventional pressure taps and controls are also unsatisfactory in this situation as the pressure within the pipeline is insufficient to raise the jumper valve and washer within the tap. For this reason, plug-type taps are used or, alternatively, the jumper valves may be permanently attached to the spindles if a more modern style of tap is required.

The elevated stands supporting water storage tanks should be carefully designed and manufactured. Tank suppliers are usually able to provide details of materials, sizes and methods of construction. Ready-made stands

are also available. When galvanised steel tanks are to be placed on elevated stands it is essential to attach the tank to the stand platform so that the tank will not be blown off the stand by strong winds when the tank is empty. Figure 8.7 shows the installation of a concrete tank which will have to be partly filled to prevent it from 'floating' under wet conditions. Figure 8.8 shows common methods of holding down storage tanks.

Pressure system

The pressure system has increased in popularity due to the improvement in the lifestyle of people in rural and

FIG 8.9 Appliance rating chart

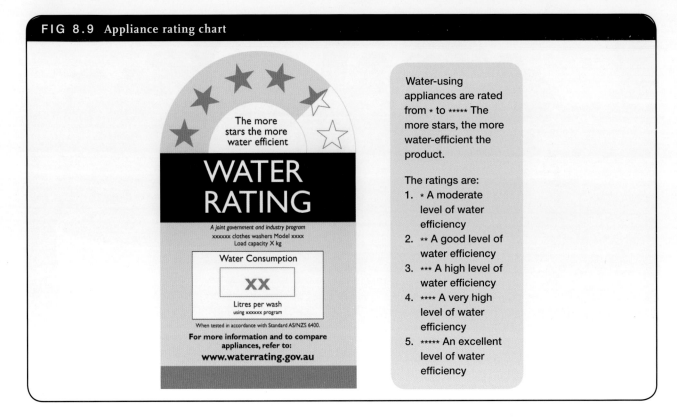

The more stars the more water efficient

WATER RATING

A joint government and industry program
xxxxxx clothes washers Model xxxx
Load capacity X kg

Water Consumption

XX

Litres per wash
using xxxxxx program

When tested in accordance with Standard AS/NZS 6400.

For more information and to compare appliances, refer to:
www.waterrating.gov.au

Water-using appliances are rated from ∗ to ∗∗∗∗∗ The more stars, the more water-efficient the product.

The ratings are:
1. ∗ A moderate level of water efficiency
2. ∗∗ A good level of water efficiency
3. ∗∗∗ A high level of water efficiency
4. ∗∗∗∗ A very high level of water efficiency
5. ∗∗∗∗∗ An excellent level of water efficiency

isolated communities. It is possible to equip most modern homes in these areas with pressurised hot water systems, automatic washing machines and dishwashers, which all rely on water being supplied at pressure for their operation.

Water storage for these systems may be situated either above or below ground and the water is pressurised in the system by means of a close-coupled centrifugal pump. The required pressure is controlled by a pressure switch activated by a pressure vessel attached to the pump. For details of this equipment see Chapter 5, Valves, taps and controls. When this type of system is used, control taps designed for reticulated supplies may be used without modification.

WELS WATER RATING

Labelling scheme

It is paramount in rural areas that water conservation is adhered to on properties and townships.

The National Water Efficiency Labelling Standards Scheme (WELS) enables consumers to identify and select water-efficient products (Figure 8.9). It is administered by the Federal Government.

The products currently covered are:

- shower heads
- dishwashers
- clothes washing machines
- urinal operating mechanisms
- taps and tap outlets
- toilet suites
- flow regulators.

NOTE: The compliance requirements for the Water Conservation Rating and Labelling Scheme are specified in Standards Australia publication *Manual of Assessment Procedures for Water Efficient Products.* (SAA MP64)

FOR STUDENT RESEARCH

Research the water quality of the Great Artesian Basin.

Australian Standards

AS/NZS 3500.1 2003 Plumbing and drainage—Water services
AS/NZS 4020 Testing of products for use in contact with drinking water
AS 2179 Metal rainwater goods—Specification
AS/NZS 4766 Polyethylene storage tanks for water and chemicals
HB 230-2008 Rainwater tank design and installation handbook

CHAPTER NINE

Pumps and pumping

LEARNING OBJECTIVES

In this chapter you will learn about:

9.1 pump terminology

9.2 pumps and their application

9.3 pump installation

9.4 pressure systems.

INTRODUCTION

Where the available water supply cannot meet the minimum required pressure and flow rates, storage tanks and/or pumps can be installed to achieve the desired demands. Pumps are designed to transfer liquids to the required elevation. They come in various sizes, types and shapes for different applications such as raising and/or pressurising water, raising sump water and the circulating of fluids.

TERMINOLOGY

Before discussing pumps and pumping, it is necessary to understand some of the terminology commonly used when referring to the operation of a pump.

A *pump* may be defined in several ways but for the purpose of this text it is best defined as: a device capable of supplying energy to a liquid, causing the liquid to move from one place to another. The *static head* is the vertical head in metres from the surface of the liquid to be pumped to the point of discharge. This distance is usually divided into two separate static heads:

1 *Static suction head* This is the vertical distance from the liquid to be pumped to the centre of the pump inlet (Figure 9.1 (a)).

2 *Static delivery head* This is the vertical distance from the centre of the pump outlet to the point of discharge (Figure 9.1 (b)).

It is necessary to separate these two static heads as pumps of various designs have critical limitations, especially on the section side of the pump, and when

Example: A farmer wishes to irrigate a paddock using water from a dam. The vertical distance from the water level in the dam to the inlet of the pump is 2 m (suction head) and the vertical distance from the pump outlet to the sprinkler head is 10 m. The pressure required to operate the sprinkler effectively is 150 kPa. The method of calculation for the static and pressure heads against which the pump is required to operate is shown in Figure 9.2.

specifying a pump for a particular installation it is important that these limitations are not exceeded.

Total head

This is made up of three separate components:

1 static head (suction and delivery) (Figure 9.1(c))

2 pressure head

3 frictional and velocity head losses.

Pressure head

We have assumed so far that the surface of the liquid on the suction side of the pump is open to the atmosphere and that the delivery side is discharging into an open storage tank. This is not always the case and quite often a pump is required to deliver liquids at pressure. To achieve this, the pump is required not only to overcome the static head (suction and delivery) but also to supply the liquid at a given pressure. As pump performance is usually expressed as 'metres head' and not as a pressure reading, the pressure required at the outlet in kilopascals should be converted to metres of head and added to the delivery head.

Friction head

This is the head required to overcome friction that occurs between the wall of the pipe conveying the water and the water itself. There are several factors that contribute to head loss due to friction and they apply not only to pumping but to all water supply systems. The factors are:

1 the length of pipe

2 the internal surface area of the pipe

3 the smoothness of the internal surface of the pipe

4 the velocity of the water flowing through the pipe

5 the number of changes in direction and the design of fittings and controls used

6 the temperature of the liquid being pumped.

When friction occurs in the suction side of the pump it will reduce the height to which the water may be raised. The friction created on the delivery side of the pump will make pumping more difficult, placing additional strain on the pump and reducing the amount of water that may be raised to a given height.

If it were possible to design a pump capable of creating a perfect vacuum in the suction pipe, the atmospheric pressure exerted on the surface of the water to be pumped would be sufficient to force the water up the suction line approximately 10.3 m.

Example: 1 m head of water exerts a pressure of 9.81 kPa. Atmospheric pressure at sea level is approximately 101 kPa.

$$\frac{101}{9.81} = 10.2956$$

∴ atmospheric pressure could support 10.3 m head

FIG 9.1 (a) A = static suction head, (b) B = static delivery head, (c) A + B = static head

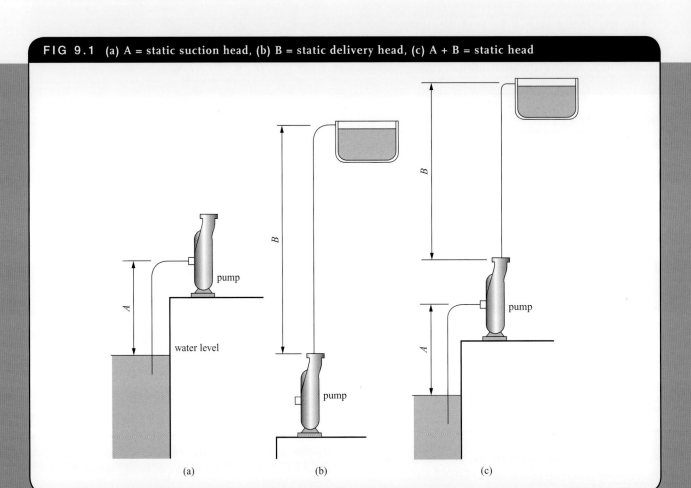

(a) (b) (c)

FIG 9.2 Static suction head A = 2 m; static delivery head B = 10 m; pressure head 150 ÷ 9.81 = 15.29 m

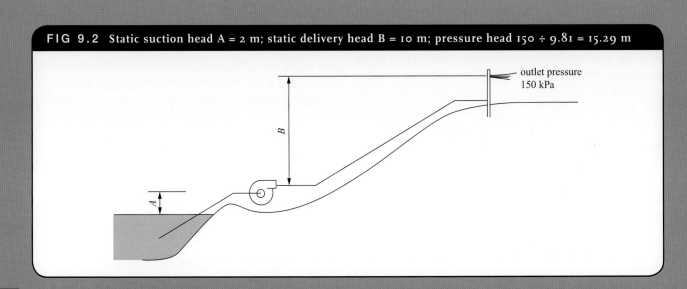

outlet pressure
150 kPa

This is purely theoretical, however, as a perfect vacuum is impossible to obtain due to slip and friction in the pump. In practice, it is found that the maximum height to which water will rise in a suction pipe is approximately 8 m.

PUMPS AND THEIR APPLICATION

Many types of pump are manufactured and each is designed for a specific pumping application. Although the plumber is often responsible for the initial installation of the pump, or at least the provision of inlet and outlet connections, the repair and maintenance of pumps is usually the responsibility of the manufacturer while the pump remains under warranty. However, as the person most directly concerned with the supply of water to properties, the plumber must understand the operation and application of the various pump types.

Pumps are generally divided into two groups:

1 positive displacement pumps

2 rotodynamic or centrifugal pumps.

Positive displacement pumps

Positive displacement pumps are usually operated by a reciprocating action that positively fills and empties a chamber with a fixed capacity, the amount of liquid moved being controlled by the speed of the reciprocating action. There are two basic types of positive displacement pumps:

1 reciprocating positive displacement pumps

2 rotary positive displacement pumps.

Reciprocating positive displacement pumps

The lift pump

The lift pump is the simplest form of reciprocating pump and consists of a piston that moves up and down inside a cylinder. A hinged or loose valve is fitted to the piston and a suction valve is fitted at the entry point of the suction pipe and the cylinder. The suction pipe is fitted with a foot valve and strainer, as are most pumps. The foot valve maintains water in the suction line on completion of pumping, reducing the need to prime the pump when pumping is recommenced. The strainer is intended to prevent foreign matter entering the valves and affecting their operation.

Figure 9.3(a) shows the commencement of the upstroke. The initial upward movement of the piston reduces the pressure in the lower part of the cylinder, allowing the pressure of the atmosphere acting on the surface of the water to force the water up the suction pipe. As the water flows up the suction pipe, the suction valve opens, allowing the water to enter the lower end of the cylinder.

FIG 9.3 Lift pump: (a) Commencement of the upstroke, (b) effect of the downstroke, (c) effect of the upstroke

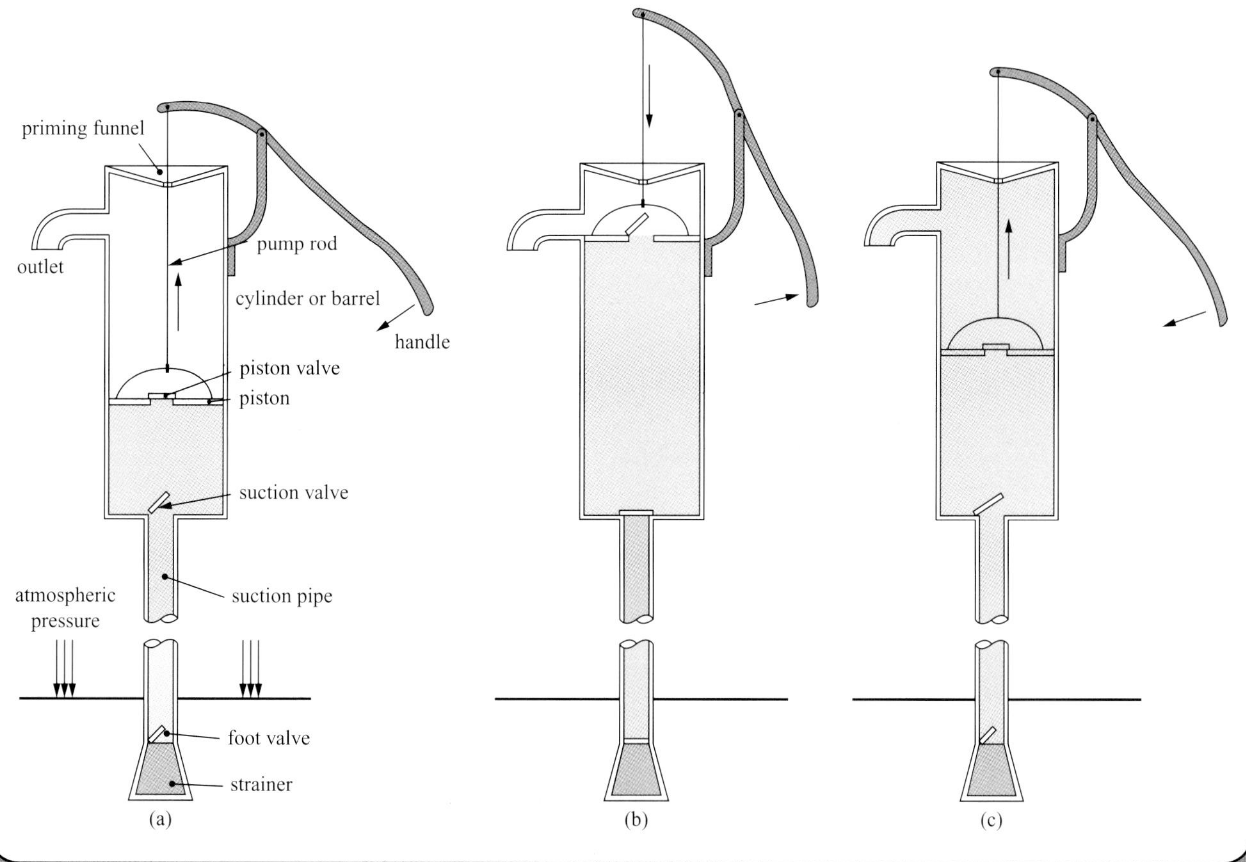

Figure 9.3(b) shows the effect of the downstroke. The downward movement of the piston closes the suction valve and opens the piston valve, allowing the water trapped in the cylinder to pass through the piston.

Figure 9.3(c) shows the effect of the next upstroke. The upward movement of the piston repeats the operation shown in Figure 9.3(a) and in so doing closes the piston valve, allowing the water above the piston to be lifted and to flow through the pump outlet. Water is discharged on each successive stroke of the pump.

This type of pump can only be used to raise *upstroke* water to the level of the pump outlet. Owing to the low efficiency of this type of pump it is only capable of lifting water approximately 7.5 m to 8.5 m when operating at normal atmospheric pressure.

The force pump

This type of pump (shown in Figure 9.4) has a solid piston with the delivery valve situated in the base of the delivery pipe.

On the initial upstroke (Figure 9.4(a)) the delivery valve closes and the inlet or suction valve in the base of the cylinder opens, allowing the cylinder to fill with water. On the downstroke (Figure 9.4(b)) the delivery valve opens and the inlet valve closes. The water held in the pump cylinder is then forced up the delivery pipe.

This type of pump is used to raise water above the level of the pump. However, for the pump to work satisfactorily it is essential that the suction pipe be kept as short as possible.

Combination lift and force pump

This type of pump (shown in Figure 9.5) combines the advantages of both the lift and force pumps, in that it has the ideal suction characteristics of the lift pump and the ability to force water above the level of the pump like the force pump. This pump is also fitted with an air vessel that produces a constant flow from the delivery pipe instead of the pulsating discharge that is common in both the force pump and the lift pump.

The operation of a lift and force pump is similar to the operation of a simple lift pump except that when water is discharged from the pump it is forced past the delivery valve and into the air vessel. If the delivery pipe and air vessel already contain water, the water entering the air vessel will compress the entrapped air in the top of the vessel. On the downward stroke the piston valve opens and the suction and delivery valves close. During this stroke the pressure in the pressure vessel is lessened, causing the entrapped air to expand, forcing water up the delivery pipe in a continuous stream. An air replacement valve is situated in the base of the air vessel to replace the air which may be absorbed in the water.

The pump cylinder is under increased pressure above the piston caused by the head of water in the delivery pipe. Thus the pump rod is required to have a packing gland to prevent water leaking past the pump rod.

Diaphragm pump

The diaphragm pump (shown in Figure 9.6) is a positive displacement pump fitted with a rubber diaphragm. This replaces the piston used in the lift or force pump. This type of pump is only suitable for pumping applications where small suction and delivery heads are encountered—this is due to the flexibility of the diaphragm, which stretches when excessive heads need to be overcome. The pump may be hand-operated or motor-driven and is ideal for pumping sludge or

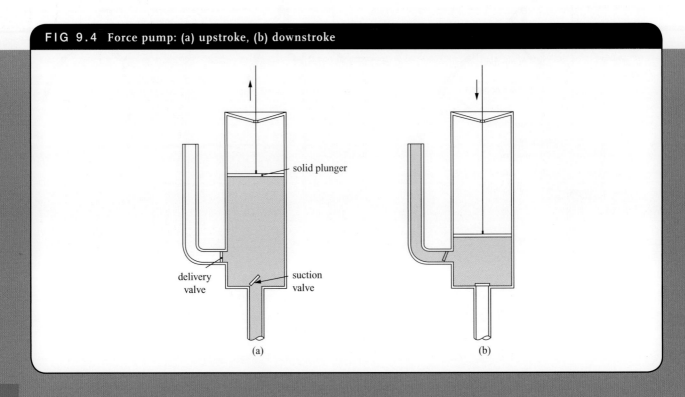

FIG 9.4 Force pump: (a) upstroke, (b) downstroke

solid plunger

delivery valve

suction valve

(a)

(b)

dewatering excavations, which may contain sand, mud and small pebbles that would damage a pump fitted with a piston.

Figure 9.6 (a) indicates the position of the diaphragm on the upstroke. As the diaphragm is lifted, a vacuum is created; this opens the inlet valve and allows water to be drawn into the pump. On the downstroke (Figure 9.6 (b)) the outlet valve is forced open and the contents of the pump are discharged through it and out of the delivery line. Water is delivered by this pump in a pulsating motion, although this may be partly eliminated when a double-acting pump is used. However, this type of pump incorporates a more complex system of valves, which increases the likelihood of the pump breaking down.

Semi-rotary pump

This is a double-acting lift and force pump which is usually hand operated (Figure 9.7). Semi-rotary pumps are fitted with two inlet valves A and B (which admit water to the pump casing) and two delivery valves X and Y (which allow the water to be transferred from the suction to the delivery side of the oscillating centre web). The oscillating web is operated by the handle. When the handle is moved backwards and forwards, suction valve A and delivery valve X operate together, as do valves B and Y.

Rotary positive displacement pumps

Rotary gear pump

This pump consists of two meshed spur gears with inlet and outlet connections situated on opposite sides of the

---

FIG 9.5 Lift and force pump: (a) upstroke, (b) downstroke

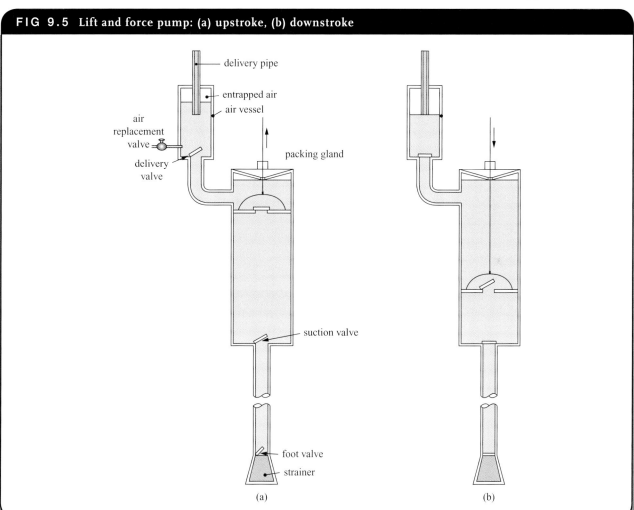

FIG 9.6 Diaphragm pump: (a) upstroke, (b) downstroke

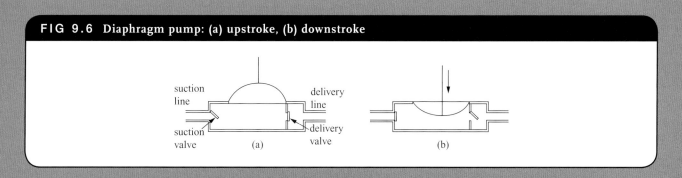

pump casing (Figure 9.8). When the pump drive shaft is rotated, the liquid to be pumped is trapped between the teeth of the gears, carried around the inside of the casing and discharged through the outlet.

The rotary gear pump is a relatively low-cost pump ideally suited to general clean-water pumping in applications where the pump is not required to run for lengthy periods. It is capable of discharging liquids at pressures up to 500 kPa. However, the higher the discharge pressure required, the greater the amount of wear that will take place in the pump.

The main disadvantage with this type of pump is that it relies on metal-to-metal contact to be effective and, if high outlet pressures are required or liquids are pumped that contain solids in suspension, the rate of wear in the pump will be greatly accelerated. For this reason, this type of pump is ideally suited to pumping oils and other liquids that can act as a lubricant.

Flexible gear pump

This pump is similar in operation to the rotary gear pump and is used in similar applications (Figure 9.9). As the flexible gear rotates, water is trapped between the teeth of the gear and the outer casing of the pump.

The outer casing is manufactured with one internal deformity so that the flexible teeth of the rotating gear are pushed over as they pass the deformity. This has the effect of squeezing out the entrapped water from between the teeth, forcing it through the pump outlet.

The main advantage of this pump over the rotary gear pump is that the metal-to-metal contact is eliminated so that the life of the pump is greatly increased.

Rotodynamic or centrifugal pumps

These pumps operate when a pressure or head is created inside a casing by the action of a spinning impeller (Figure 9.10). The impeller rotates inside a casing that is shaped to form a volute (spiral). The spinning action of the impeller causes an increase in the velocity of the water as it is thrown off the tips of the impeller into the volute. This has the effect of creating a vacuum at the centre or 'eye' of the impeller where the pump inlet is placed. Water is drawn into this vacuum by either suction or gravity, depending on the placement of the pump in relation to the source of water to be pumped (Figure 9.11).

Centrifugal pumps are used extensively in water supply work on both small and large installations and will operate successfully against very low and very high heads. In situations where the heads against which the pump is required to operate are extremely high, multi-stage centrifugal pumps may be required. These consist of a series of impellers on a common shaft revolving inside a cylindrical casing.

PUMP INSTALLATION

The most common type of centrifugal pump used on small installations is of the 'close coupled' or 'monobloc' type. This pump is permanently attached to the motor, which can be powered by either electricity or petrol. The impeller shaft is driven directly by the main shaft of the motor.

In situations where a large centrifugal pump is necessary or a mechanical advantage required, the pump and motor are mounted independently of each other, the pump being a pedestal-mounted type driven by single or multiple vee belts. This arrangement also allows

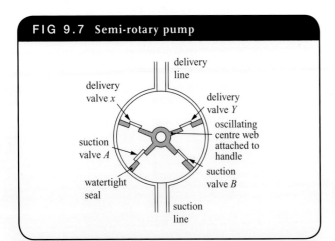

FIG 9.7 Semi-rotary pump

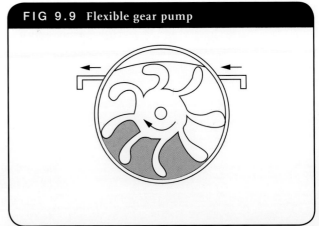

FIG 9.9 Flexible gear pump

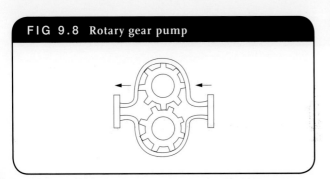

FIG 9.8 Rotary gear pump

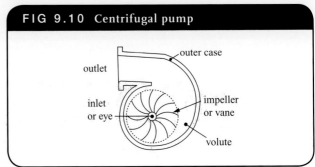

FIG 9.10 Centrifugal pump

the speed of the pump to be increased or decreased by fitting pulleys of different diameter to either the pump or motor.

Pumps and motors should be securely mounted on concrete blocks, preferably higher than the surrounding floor or ground level. This is to ensure that water leaking from packing glands or connections will not damage either the motor or the pump. This is especially important when using electric motors. Pumps should also be mounted on springs or rubber blocks to minimise vibration or noise transmission into the surrounding structure.

Pump connections

As mentioned above, the pipework connections are divided into two groups: (a) suction or inlet and (b) delivery or discharge pipework.

Suction pipework

Suction pipework should be arranged so that the suction lines are kept as short as possible and installed so that there is a gradual rise towards the pump. This is done to eliminate air pockets in the suction line, which may reduce

the suction performance of the pump. A typical air trap is shown in Figure 9.12. On installations where this type of pipework arrangement is unavoidable, air valves should be installed in positions where air is likely to be trapped.

Under normal conditions, suction pipework should have at least the same diameter as the pump inlet connection. However, suction lines are often installed using pipes of a larger diameter than the pump inlet and therefore require a reducing fitting to be installed at the inlet. In this case the traditional 'concentric' or 'equal' taper fitting should not be used as these also have a space at the top of the fitting capable of retaining air pockets (Figure 9.13). Eccentric reducers should always be used on pump connections and these fittings should always be installed with the straight side uppermost (Figure 9.14).

Suction controls

Under normal conditions there are very few controls required on the suction side of the pump and the only fittings considered essential are either a non-return valve or a foot valve with strainer (the strainer is optional). On installations that have a relatively short suction line, it is

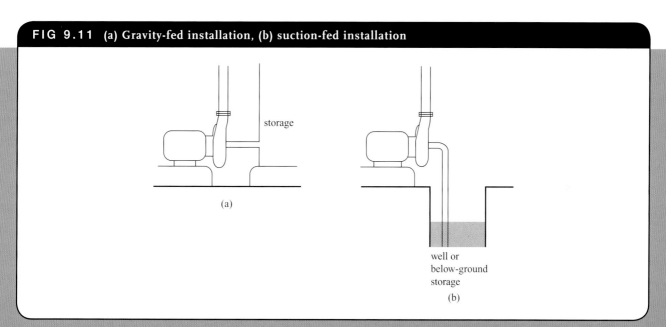

FIG 9.11 **(a) Gravity-fed installation, (b) suction-fed installation**

storage

(a)

well or below-ground storage

(b)

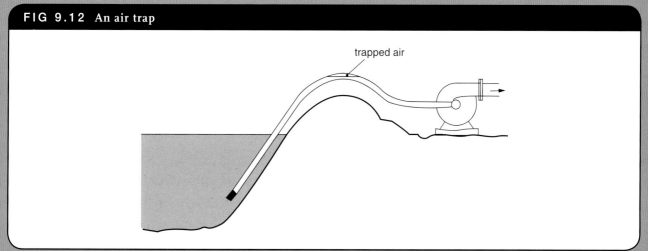

FIG 9.12 **An air trap**

trapped air

recommended practice to install a non-return valve in the line adjacent to the pump. This valve prevents the water draining from the pump casing and suction line when the pump is turned off, thus eliminating the need to prime the pump each time it is switched on. A swing type valve is the most suitable in this instance since it offers less restriction to the flow than the under-and-over type.

Where the suction line is long, the non-return valve adjacent to the pump is dispensed with and a foot valve is attached to the inlet of the suction line. This valve serves the same purpose as a non-return valve but is positioned so that the full length of the suction line may be primed. A strainer is often incorporated in this type of valve. A union and flexible pipe should be installed on both the suction and delivery pipework.

Delivery (or discharge) pipework

Delivery or discharge pipework should be designed and installed so that frictional losses are kept to a minimum. The head losses through bends, branches and controls soon mount up, and although they may be easily overcome on a short delivery line, the cumulative effect over a long distance may require an increase in pipe diameter or pump size to achieve the desired delivery.

Fittings such as bends should be fabricated with a long radius to minimise frictional losses and branches should be Y-pattern or sweep type as in Figure 9.15 (a) and (b).

Delivery controls

Controls on the delivery side of a pump should be kept to a minimum for the reasons stated above. However, it is essential that the pump installation be fitted with a non-return valve and a control valve adjacent to the pump. These valves are installed to prevent the contents of the delivery line from draining back through the pump casing when the pump is turned off and also to prevent water loss should the pump be disconnected for servicing. Automatic air bleed valves should also be installed at high points in the delivery line so that air may be exhausted from the line when pumping commences after a prolonged shutdown.

Groundwater pump restrictions

Identifying the characteristics of your water source is vital for the quality of your irrigation. Different water sources must of course be managed differently. The performance of the pump relies heavily on a systematic analysis of the water source, and making the proper selection of equipment. Surface water flow is relatively easy to understand, because it is readily observed and easily measured. Groundwater flow is, however, hidden, making measurements more complicated.

The most common restrictions to groundwater supply are:

- supply limitations
- pump wear
- clogging
- over pumping.

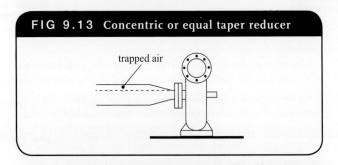

FIG 9.13 Concentric or equal taper reducer

trapped air

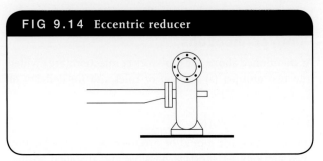

FIG 9.14 Eccentric reducer

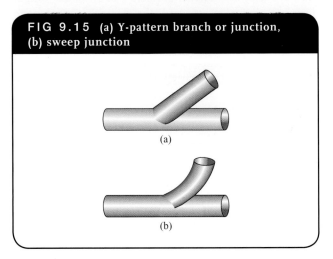

FIG 9.15 (a) Y-pattern branch or junction, (b) sweep junction

(a)

(b)

Supply limitations

Over-pumping a well will eventually result in dry-running, which can result in serious damage to the pump. The resulting downtime is expensive, both regarding repair costs and lost productivity. To protect the pump system from dry-running, it is extremely important to analyse how much water the well can supply.

Pump wear

An incorrect choice of pump material and the resulting pump wear (Figure 9.16) is a common problem reducing well capacity. Choosing the correct pumps with vital components made from bronze or stainless steel (Figure 9.17) from the beginning will secure a reliable, energy-efficient, and virtually maintenance-free groundwater pump solution.

Rust on cast iron pumps is created by iron from the impeller, which oxidises through contact with the oxygen in the water. When the impeller rotates, the rapidly-flowing water (5-15 m/sec) removes rust from the impeller surface. This corrosion/erosion process leads to the loss of

SUSTAINABLE PLUMBING: **GROUNDWATER**

Groundwater provides a significant source of water supplies for irrigation worldwide. It is possibly the most reliable water source we have. However, it is important to use groundwater wisely. We must ensure future water supplies and protect the fragile environment in which we live.

In remote locations, such as cattle and sheep properties, water supply can be scarce and power supply non-existent or unreliable. Renewable energy sources such as solar and wind turbines are used to power the custom-designed water pumped systems.

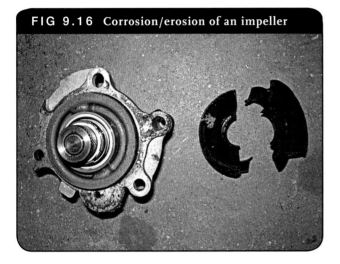

FIG 9.16 Corrosion/erosion of an impeller

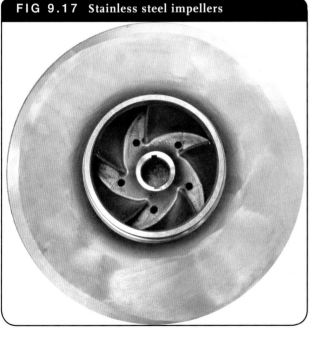

FIG 9.17 Stainless steel impellers

FIG 9.18 Sand intrusion

A sand cyclone or bag filter can be used to prevent sand, silt and rust from entering the piping system (Figure 9.18). Or an open resource/pond can be used when particle size is too small to be retained by cyclones and bag filters. The silt falls to the bottom, and irrigation water is removed from the top.

Over-pumping

Sometimes peak demand capacity causes over-pumping of the well to the level of sand intrusion. The damage can be avoided by installing a sand separator or a telescope-inserted filter section, which will reduce the quantity of silt and sand in the water.

Starting a groundwater pump when the aquifer volume is full will also result in excessive performance during the first seconds of operation (Figure 9.19). This high-capacity kick starting lifts up and releases the sand and silt in the aquifer, drawing it into the pump, but this powerful suction is eliminated by installing a 3-second ramp soft start/stop.

impeller material. When impeller material is lost, capacity and efficiency also fall.

Clogging

Piping that is partially filled with sand, silt or rust may cause some of the following problems:

- excessive power consumption
- insufficient water capacity
- pump wear.

PRESSURE (OR HYDROPNEUMATIC) SYSTEMS

In situations where the water supply pressure is inadequate or the height to which water is to be delivered is very great, such as to a very tall building, a pressure pump needs to be installed to ensure satisfactory water pressure to operate appliances and overcome frictional losses in the delivery pipework. Water pressure can be boosted by installing a pump, usually centrifugal, and a pressure-activated switch. However, this method has some disadvantages in that the pump would be required to run continuously or, it would continually cut in and out when operated by a pressure-sensitive switch. Both these methods are impractical as they place excessive strain on the motor and pump, and are generally uneconomical.

The most common method of overcoming these problems is to install a hydropneumatic system, which uses energy stored in compressed air to supply water at a predetermined pressure between upper and lower pressure limits.

The system comprises a centrifugal pump and motor unit, either direct or remote coupled, depending on the size of the installation, a hydropneumatic storage cylinder and a pressure-activated switch device. The hydropneumatic storage cylinder, as the name implies, stores both water and air and varies in capacity according to the size of the installation and the amount of water to be stored. On large installations, a unit similar to that shown in Figure 9.20 is required.

With this type of system, the water and air cushion stored in the cylinder come into direct contact causing the air to be gradually absorbed into the water. When this occurs

FIG 9.19 Groundwater bore with a submersible pump

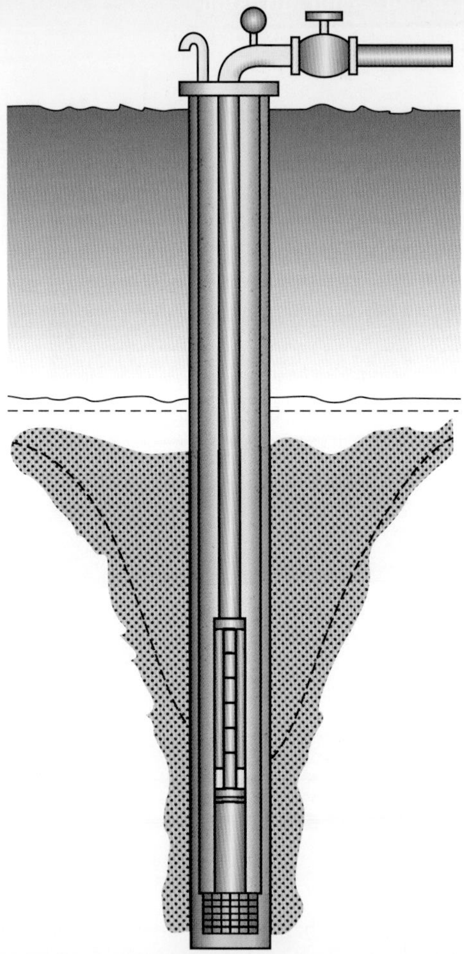

FIG 9.21 Rubber diaphragm in domestic unit

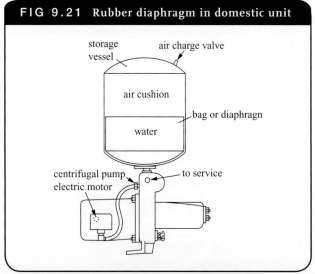

storage vessel

air charge valve

air cushion

bag or diaphragn

water

centrifugal pump electric motor

to service

FIG 9.20 Installation incorporating a large hydropneumatic storage cylinder

safety valve

pressure switch to pumps

compressed air

water under pressure

float switch operated compressor

drain

to upper floors

to lower floors

direct off mains supply

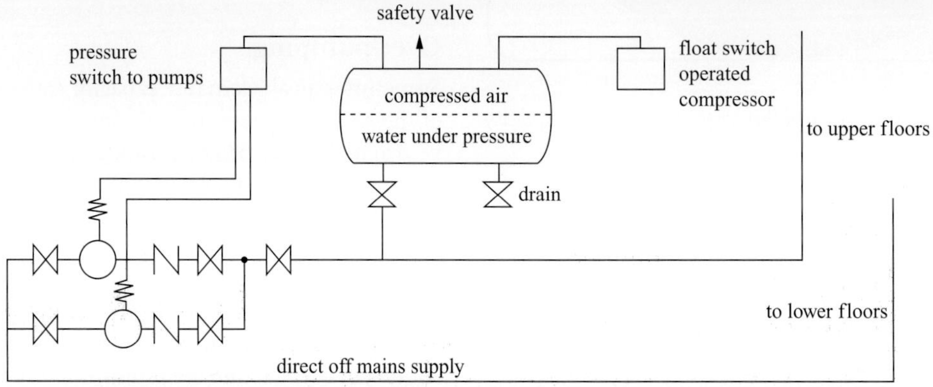

the water level rises in the cylinder, reducing the volume of the air cushion. This increases the frequency of operation of the pump as the pressure energy retained in the air cushion is reduced. To overcome this problem an air compressor is required to replace the air cushion as it is absorbed by the water, the compressor being activated by a float switch.

On smaller packaged domestic units this problem is overcome by segregating the air cushion and stored water with a rubber diaphragm or bag (Figure 9.21). Variable speed control also assists in the varying of flow to suit demand.

Sequence of operation

The operation of a system is shown in Figure 9.22. A typical hydropneumatic shallow well installation is shown in Figure 9.23.

Hydraulic ram

The hydraulic ram is a simple device that uses the kinetic energy of water falling through a relatively low height to raise a smaller amount of that water to a greater height.

The ram is only suitable for use where an unlimited supply of water is available and in situations where the ram can be installed below the water supply, since the water is supplied to the ram via the drive pipe by gravity. A sectioned diagrammatic sketch is shown in Figure 9.24 indicating the main components of the ram.

Installation

The hydraulic ram must always be installed below the level of the stored water, so that the water passing through the ram is able to free fall down the drive pipe (Figure 9.25). The excess water required to operate the ram passes through the dash valve and continues on down the creek bed.

Operation

As the water flows down the drive pipe, the velocity of the water increases until the force of water overcomes the weight of the dash valve. At this point the dash valve slams shut, stopping the flow of water through the ram instantly. The rapid closing of the dash valve converts the kinetic energy into pressure energy (water hammer), which is

FIG 9.22 Operation of a system

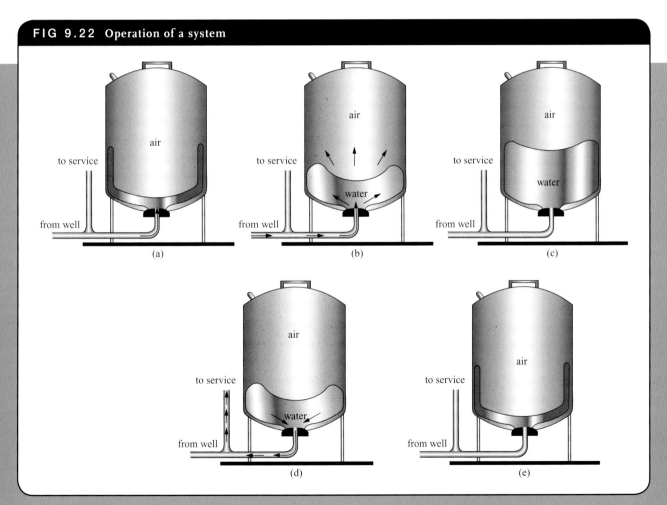

(a) The tank just after the pump starts. Water is about to enter the tank. (b) Water entering the tank. The air above the vinyl bag is compressing and the bag is filling with water. (c) When cut-off pressure is reached, the air above the bag is completely compressed and the bag is completely full of water. (d) When water is drawn from the tank, the compressed air forces the water out of the bag. (e) This continues until the bag is completely empty and the cycle is ready to start again.

FIG 9.23 Typical installation of a hydropneumatic shallow well

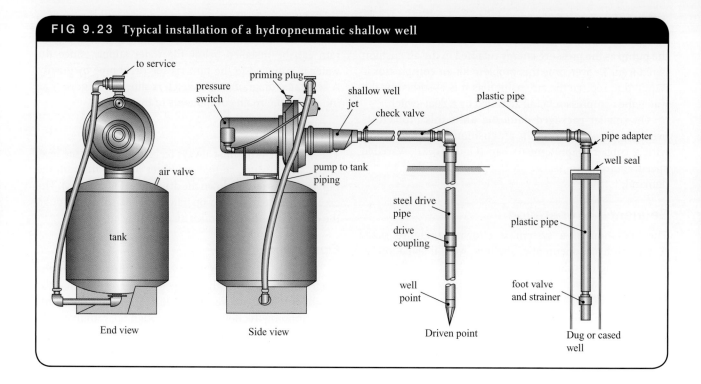

End view

Side view

Driven point

Dug or cased well

FIG 9.24 Main components of hydraulic ram

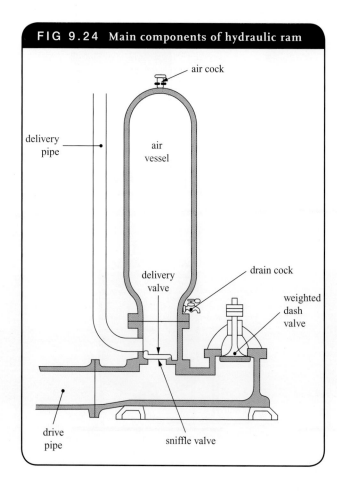

pipe where the surplus energy is dissipated. When the pressure is reduced inside the ram, the dash valve drops open and the delivery valve closes, allowing the water to flow down the drive pipe again and the cycle of operation is repeated.

It is important to ensure that the drive pipe is of sufficient length and diameter so that sufficient water is supplied to the ram. By providing a drive pipe that has a minimum length of 15 m and is at least twice the diameter of the delivery pipe, the tendency of the dash valve to operate slowly should be eliminated. Drive pipes should be kept as straight as possible with an even fall over their entire length. They should be fitted with a suitable strainer at the inlet to prevent the entry of foreign matter into the ram.

The air vessel should be of sufficient size to dampen the shock caused by the closing of the dash valve. When the dash valve closes and the delivery valve opens, the air in the top of the vessel is compressed by the incoming water. As the water in the ram recoils up the drive pipe, the delivery valve closes and the previously compressed air expands and forces the water up the delivery pipe to the storage tank.

As the air in the air vessel comes into direct contact with the water under pressure, it is gradually absorbed. For this reason, the air vessel should be recharged with air at frequent intervals. To prevent the air cushion being absorbed too quickly, some rams are fitted with a 'sniffle' valve, which allows a small quantity of air to be drawn into the vessel at each operation of the dash valve.

Hydraulic rams, if adjusted correctly are capable of operating successfully with a fall in the drive pipe as little as 600 mm. However, it is considered good practice to design the system so that the fall in the drive pipe is not less than 2 m.

sufficient to open the delivery valve, thus allowing a small volume of water into the air vessel and up the delivery pipe. The resistance offered by the delivery valve is sufficient to cause the remainder of the water to 'recoil' up the drive

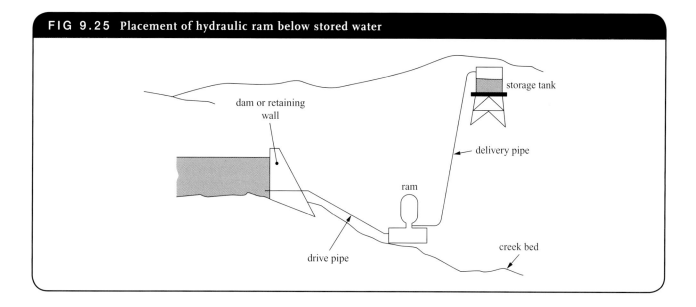

FIG 9.25   Placement of hydraulic ram below stored water

The height to which water may be delivered from a ram varies considerably and is in proportion to the head operating on the ram. For example, an average ram is capable of delivering a quarter of the drive water to three-and-a-half times the height or head operating on the inlet, one-eighth to six times the head and one-sixteenth to ten times the head. It is evident from these figures that the efficiency of the ram decreases rapidly as the delivery head increases.

FOR STUDENT RESEARCH

Working with your supervisor, create a project, with the aid of photos and brief explanations, of all the pumps that are used within the campus.

Australian Standards

AS/NZS 3500.1: 2003 Plumbing and drainage—Water services

ON-SITE STORIES 9.1

OOPS...

Wayne Clayton, *Plumbing Lecturer, Polytechnic West, WA*

I got a call from an excavator operator mate of mine who said that the pump well from his septic tank was overflowing and could I take a look at it. I organised a replacement pump to be delivered and on arrival I found the pump well was indeed overflowing. I told my mate that the pump had probably failed and it would need to be replaced. We discussed the possible causes of its failure as I removed the access lid. As I did, condoms floated to the surface and out of the pump well into the yard.

Now, being a man of the world, I resisted the urge to joke about it. Instead I said, 'That will be the cause of the pump failure. As the well is pumped down, the condoms have been sucked into the impeller and seized the electric motor, burning it out.'

I hadn't realised that he had not said a word for a while.

'I would advise you not to dispose of them down the toilet in future,' I piped up.

With that, he looked at me and said *'They're not bloody mine.'* OOPS!

I replaced the pump and left before his wife got home.

Glossary

A

access opening an opening in a building element, fitted with a removable cover, to allow maintenance of a concealed pipe, fixture or other apparatus

acetone an inflammable and volatile liquid used as a solvent in acetylene cylinders to dissolve and stabilise acetylene under pressure

acetylene a highly combustible gas composed of carbon and hydrogen (C_2H_2) used as a fuel gas in oxyacetylene welding and cutting. When burned with oxygen in the correct proportions it produces a flame temperature of approximately 3000° C

act legislation passed by parliament to control a specific area of undertaking, e.g. *Sewerage Act*

aeration systems systems designed to treat liquid waste by the processes of air injection

air admittance valve a component fitted to a plumbing system allowing air to enter the piping system, but not allowing air or gases out

air eliminator a device which opens to release accumulated air from a hydraulic system and which automatically closes in the presence of a liquid

air gap 1. (sanitary plumbing) the unobstructed vertical distance through the free atmosphere between the outlet of waste and the overflow level of the receptacle into which it is discharging

2. (water supply) the unobstructed vertical distance through the free atmosphere between the lowest opening from any pipe or fitting supplying water to a tank, fixture or other sanitary fitting and the spill level of the receptacle. The gap is to prevent backflow of water from the cistern or tank into the supply line

air test a test of the constructed performance quality of pipelines, where air loss of pressurised air is observed either directly, by a soap solution, or indirectly by comparing the measured rate loss of air pressure against predetermined acceptable limits

alignment the setting in a straight line of a number of points, e.g. pipe lengths in a pipeline

annealing the process of gradually cooling a metal part after welding, or reheating it to make it soft enough for mechanical working. Annealing will relieve stresses in an existing metal or stresses that may be set up by welding operations

approved approved as directed by the relevant water, sewerage or drainage authority

aspect the direction something faces, e.g. a northerly aspect

atmospheric pressure the pressure exerted by the atmosphere at a given point, e.g. sea level

Australian Standards approved standard for materials, equipment, techniques or procedures as set down by the Standards Association of Australia

authority the appropriate body authorised by statute to exercise jurisdiction over the installation of plumbing, gasfitting, sewerage and drainage works

B

backfill material used for filling trenches and excavations after a pipeline has been laid

backflow a flow of water in the reverse direction to that intended

backsight (BS) the first reading taken on a levelling staff and logged in the appropriate column in the field book after the initial setting up of the levelling instrument

basement the floor space of a building, which is substantially or wholly below the level of the adjoining ground surface

batten square or rectangular timbers fixed to rafters so that a roof covering may be attached to them

bedding material beneath and 'cradling' a pipe or drain

bench mark (BM) stable reference points, the elevations of which have been accurately determined. Bench marks are taken as permanent reference points during levelling operations. Temporary bench marks (TBM) are used as interim marks

boss a pipe or fitting formed as an integral part of a tank or vessel to facilitate the connection of inlet and outlet pipework

boundary trap a composite fitting incorporating a trap for the prevention of the passage of gases from the sewer to the drain and the passage of prohibited substances

branch the intersection of two pipes

branch drain any branch off a main property drain

branch pipe a discharge pipe to which two or more fixture traps at any one floor level are connected

branch vent a graded vent at any one floor level interconnecting two or more individual trap or group vents

breather vent a DN50 vent from the topmost junction of a sealed gully riser to the atmosphere, extending a minimum of 300 mm above the lowest fixture discharging to the sealed gully

braze welding a joining process that, unlike brazing, does not depend on capillary attraction. The parent metal is not melted but the joint design is similar to that which would be used if a fusion weld were made. The filler metal is generally a non-ferrous metal or alloy with a melting point above 500° C

brazing a joining process in which the molten filler metal is drawn by capillary action between two closely adjacent surfaces. The filler metal is a non-ferrous metal or alloy with a melting point above 500°C but lower man that of the metal being joined

breaktank an intermediate tank to reduce the operating head on an outlet

BSP thread British Standard Pipe thread

building any building used as a workplace, residence, place of business, entertainment institution, place of human habitation or as a factory, which contains plumbing fixtures

butt weld a weld in which the two edges of the pipe material to be united are butted together.

bylaws regulations made by local and other authorities under relevant legislation

C

capacity (of a cistern) the volume calculated by measuring up to the marked water level

capacity (of a storage tank) the volume of water above the invert of the outlet pipe when the water surface in the tank is 20mm below the overflow level

capillary action the force of adhesion existing between a solid and a liquid in capillarity—occurs between two close-fitting smooth surfaces

capillary joint a joint in which the parts are united by the flow of filler metal by capillarity along the annular space between the outside of the tube and the inside of the fitting

carburising flame a mixture of oxygen and acetylene gas in which mere is an excess of acetylene

catchment area the area of land from which run-off is taken to fill a dam or reservoir

caulking rendering a joint watertight by compressing the sealing material

ceiling joist a timber member fixed to wall plates to enable ceiling linings to be fixed

chalk line string or twine containing chalk dust, The line is drawn tight then plucked to produce a chalk line on the work surface

chase a recess cut into brickwork to allow for piping etc.

cistern a storage tank in which the water is at atmospheric pressure. The water is usually received through a float control valve set, at a predetermined level and incorporating an air gap

coagulant a substance which when added to other substances causes them to congeal or coagulate

coagulation the joining together or congealing of substances into semisolid mass

colloid an apparently soluble substance which can be strained from a liquid

combined soil and waste pipe any pipe which receives the discharge from both the soil and waste fixtures and conveys them to the drain

common rafter the rafter which runs from the fascia to the ridge

common vent a vent installed at any floor level. It is provided for venting the traps of not more than two fixtures individually connected, and is usually a vertical extension of a graded pipe or branch

compaction the process of consolidating backfill by mechanical or other means

compression joint a joint made by fittings in which the end of the pipe is held under compression:
1. *manipulative* a fitting in which the joint is made by compression of a ring or sleeve or part of the fitting on the outside wall of the tube
2. *non-manipulative* a fitting in which the direction of compression is through the axis of symmetry of the ring

conduit 1. a pipe or channel usually of large size used for the conveyance of liquids
2. a pipe of large diameter through which a smaller pipe passes

connection that part of the drain which connects a property drain with the main sewer

contour line a line that indicates points of equal height, generally spaced at one metre intervals on construction drawings but may be greater distances apart on maps

cross-connection faulty plumbing design which may permit the entry of contaminated water into a drinking supply

cross-vent a vent which interconnects a stack with its relief vent

cylinder a portable steel or aluminium container for storing and transporting industrial gases such as compressed oxygen or dissolved acetylene

D

downstream vent a drainage vent located on, or connected to a drain discharging to a boundary trap

datum plane (or **datum**) a horizontal plane of known height to which elevations of different points can be referred. The mean sea level is the level surface generally adopted as a datum. The Australian Height Datum (AHD) is based on the mean sea level determined by the tide gauge readings around the Australian coast

depth of fusion a welding term expressing the distance that the fusion extends into the base material

dewpoint the temperature at which water begins to form from vapour

dezincification the selective corrosion of copper alloys (brasses) in which the alloy loses its zinc component and is converted into a porous shell of copper which has poor mechanical strength

diameter straight line passing through the centre and from one side to the other of a circle or solid circular figure

dimensions measurements showing the size and extent of an object

discharge pipe any pipe for the conveyance of sewage or trade waste

domestic fixture a fixture or appliance which is designed for use in residential situations only. A fixture or appliance of this type may be installed in a non-residential building, but the waste which it discharges must be similar to that of its domestic counterpart

downstream towards a lower level

drain the line of pipes, normally laid underground including all fittings for the conveyance of sewage and/or trade wastes to the sewer

duct an enclosed area to accommodate pipework

E

eaves overhanging edges of roofs

eaves lining a lining used to close in the bottom of an overhanging edge of a roof. *See* soffit

effluent the liquid contents of a septic tank or other waste treatment plant which is discharged after a reasonable period of standing

electrolysis corrosion corrosion produced by the contact of two dissimilar metals in the presence of an electrolyte

elevation level *see* reduced level (RL)

evaporation conversion of a liquid to the vapour state by the addition of latent heat

expansion coupling a coupling that permits longitudinal movement of the joined parts, caused by expansion and contraction due to temperature change

expansive soil soil that undergoes volumetric changes due to variations in moisture content

F

fall the difference in elevation between two given points (e.g. on a gutter, which allows water to flow to the downpipe)

filtration the process of removing finely divided solids and floc from water by passing it through beds of specially selected and graded sands and /or filters

fitting any component of a sanitary plumbing installation other than a fixture or pipe

fixture a device, the operation of which results in a discharge into the sanitary plumbing installation

fixture discharge pipe the pipe to which the single fixture trap is connected

fixture trap a trap connected directly beneath the outlet of a fixture

fixture unit a unit of measure based on the rate of discharge, time of operation and frequency of use of a fixture that expresses the hydraulic load imposed on the sanitary plumbing installation

fixture unit rating the system loading value in fixture units assigned to a fixture which in turn is used to determine the size of traps, pipes and vents within the system

flame (excess acetylene) *see* carburising flame

flame (excess oxygen) *see* oxidising flame

flanges raised flat-faced fittings attached to the ends of pipes and connected together by bolts

flashback the burning back of the flame into the blowpipe, or the ignition of an explosive mixture in one of the gas lines

flocculation the process brought about by stirring a coagulant into raw water, resulting in minute gelatine-like particles (floc) forming around the turbidity-producing substances

floor waste grated inlet within a graded floor, intended to drain the floor

floor waste gully a disconnector gully for installation inside a building, for use with a floor grating or waste outlet fitting on a riser pipe and with provision, where required, for connection of waste pipes for fixtures

fluoridation the process of adding minute doses of fluoride compound to water at approximately one part per million. The purpose of this process is to reduce dental cavities in the population

flux a chemical powder or paste used to dissolve oxides, clean weld metal of undesirable inclusions and prevent oxidation of a metal during welding or brazing operations

foresight (FS) the last recorded staff reading for a levelling position. It is logged in the appropriate foresight column in the book before moving the levelling instrument to a new position

fully vented system a system of plumbing with provision for the separate ventilation of every fixture trap connected other than to a floor waste gully, and of the trap of every floor waste gully

fully vented system (modified) a system of plumbing which differs from a fully vented system in that the traps of any group of two or more fixtures or floor waste gullies, discharging to the same branch pipe, are vented in common by one or more group vents connected to such pipe

G

gable triangular upper section of a wall at the end of a ridged roof

galvanising a zinc coating used on steel to prevent rusting

girt a steel purlin or structural member

grade the angle of inclination expressed as the ratio of unit rise to horizontal distance or as a percentage, i.e. 1:50 or 2 per cent

graded pipe a pipe or drain installed at a grade

graphics the art of making drawings, especially in mechanics, in accordance with mathematical rules

graphite a type of carbon in mineral form, used as a lubricant

grating a framework of metal strips fitted over the inlet of waste outlets and gully traps, to prevent the ingress of large solids

ground the surface of earth, soil or rock which conforms to the established finished grade at a specific location after all excavations have been backfilled and all surface treatment completed

groundwater water occurring in the subsoil

group vent a vent connected to a branch to which unvented fixtures are connected

gully an assembly used in a waste outlet system comprising a trap and fittings:

1. *disconnector gully (external)* located outside a building fitted with grate and finishing collar or yard sink, providing disconnection of waste and soil

2. *disconnector gully (sealed)* a gully normally situated inside a building sealed with a removable airtight top and breather ventilator, providing the same function as 1

3. *overflow relief gully (ORG)* a gully fitted with a loose grating, used as a surcharge relief in the event of a sewage overflow

4. *floor waste gully* a gully positioned inside a building, provided with a floor grate or waste outlet fitting, and with provision for the connection of waste discharges to its riser. The diameter of such a gully would seldom exceed 80 mm

H

hatching method of shading with multiple lines; often used to show a cross-section through a solid object

head vent the vertical pipe which is the continuation of a drain at its upper end

header vent the vent interconnecting the tops of two or more relief vents or stack vents

heat a form of energy capable of performing work

heel the outside of a bend

height of collimation (HC) sometimes referred to as 'height of instrument', is the imaginary line passing through the intersection of cross hairs and the optical centre of the object lens. This is the horizontal line to which all staff readings are taken. It is important to note that the collimation line is only horizontal when an instrument is in perfect adjustment and is set up and levelled

hip the external intersection of two pitched surfaces on a roof

hose *see* tubing

hot-dip galvanising a process by which iron or steel is immersed in molten zinc to provide protection against corrosion

hot discharge a discharge at a temperature of 40°C or higher

hydraulics the branch of science and technology concerned with the mechanics of fluids, especially liquids. The study of liquids in motion

hydrology the science that treats the occurrence, circulation, distribution and properties of water and its reaction with the environment

hydrostatic test a test that subjects a pipeline to a static head of water pressure

hydrostatics the study of liquids at rest and the forces exerted on them or by them

I

impact a force exerted when one body collides with another

indirect connection when a fixture waste discharges above a water-seal

inspection opening an access opening in a pipe or drain sealed with a removable plug or cover used as an access for the purposes of inspection, maintenance and hydraulic testing

insulation a material which reduces the transmission of heat, sound, electricity or moisture

intermediate sight (IS) all readings recorded between a backsight and a foresight

invert the lowest point of the internal surface of a pipe at any cross-section

isometric drawing a method of non-perspective pictorial drawing in which the object being drawn is turned so that three mutually perpendicular areas are equally foreshortened. All dimension and projection lines are at 30° to the horizontal

J

jointing the lapping of two pieces of material so that they can be joined by soldering, riveting or other approved methods

L

lagging *see* insulation

laser a fine beam of light

latent heat (also known as hidden heat) the amount of heat required to change the state of a substance without changing the temperature, e.g. ice at 0° C to water at 0° C

leeward the side away from the direction from which the wind is blowing

long bend a pipe bend greater than 45° C having a centre-line radius equal to or greater than 1.5 times the diameter

M

maceration wasting away by steeping; separation of parts

macerator pump a positive displacement pump containing a macerating device

main the principal pipeline in a water or drainage reticulation system

main drain that drain which determines the depth of the connection

mandatory required by law

mitre bend a pipe bend made with the use of a mitre cut (angle cut)

N

neutral flame an oxy-fuel gas flame in which the inner cone, or that portion of the flame used, is neither oxidising nor carburising. It is characterised by an almost colourless outer envelope and a sharply defined inner cone without feather or secondary flame

neutraliser a device for neutralising acid waste water prior to being discharged into the drains

nominal size standard sizes of pipes and fittings in accordance with the relevant Australian Standards

O

offset the pipes and fittings used to provide continuity between pipes of parallel axis but which are not on line

operative plumber a plumber who works under the supervision of a licensed plumber

orthogonal drawing method of producing views of an object by projecting straight perpendicular lines from that object to a viewing plane

orthographic drawing *see* orthogonal drawing

osmosis diffusion of liquids through a porous layer of skin

outlet (nozzle, pop, spout) an opening in a sanitary fixture, appliance or vessel serving to discharge the contents

overflow level the level of the rim of a fixture, or the invert level of an overflow pipe

oxidising flame an oxy-fuel gas flame in which the inner cone, or that portion of the flame used, has an excess of oxygen. It is characterised by the length of the inner cone when compared with the cone of a neutral flame

oxygen a colourless and odourless gas which supports combustion and is present in the atmosphere to the extent of approximately 21 per cent by volume. When the correct mixture of oxygen and acetylene is burned, a flame temperature of approximately 3000°C is obtained

P

pan connector a plastic or rubber fitting for connecting a WC pan or soil fixture to a soil pipe

pH an indication of water's acidity or alkalinity. The scale is ranked from 0 (acid) to 14 (alkaline)

pH correction the process of adjusting the acid/alkali balance of water

pipe support a device for supporting and securing piping to walls, ceilings, floors or structural members

plumb-bob a pointed weight suspended at the end of a string line to test for perpendicularity

plumbing system fixtures, fittings, pipes, materials, or appliances other than the sanitary drain, used for the collection and conveyance of any wastes or waste waters from any premises and includes all vents, flashing and water service connected to fixtures

polyvinyl chloride (PVC) polymer of vinyl chloride; tasteless, odourless and insoluble in most organic solvents. A member of the family of vinyl resins and used for a wide range of pipes and fittings

potable water water suitable for drinking, culinary and domestic purposes

p-trap a trap constructed with the inlet leg vertical and the outlet leg inclined below the horizontal within specified limits

pressure ratio valve a valve that automatically reduces outlet water pressure to below a specified ratio of its inlet pressure

pressure reducing valve a valve that automatically reduces the pressure to below a predetermined value on the down steam side of the valve

profile the shape of something when viewed from the side

prohibited discharge a waste considered by the water authorities to be dangerous to the disposal system or to the employees at a treatment works

projection the system of producing lines on a flat surface from features that may be flat or curved

pump a mechanical device generally driven by a prime mover, and used for raising fluids from a lower to a higher level

PVC-U unplasticised polyvinyl chloride. *See* polyvinyl chloride

R

radiant energy heat transmitted from one body to another without heating the intervening medium

radius a straight line from the centre to the circumference of a circle

reduced level (RL) the height or elevation above a given point, adopted as a datum

regulator a device for controlling the delivery of gases at a chosen constant pressure regardless of variation in cylinder or pipeline pressure

relative humidity the ratio of the amount of moisture held in suspension in the air at any temperature to that which could be held in the air at the same temperature expressed as a percentage

relief vent a vent branching from a stack below the point of connection of the lowest fixture discharge

ridge the highest point of a roof; occurs at the apex where the common rafters meet

rise the amount by which a given point is higher than the previous point

riser a straight length of pipe fitted to the inlet of a trap and extending to floor level, fitted with approved grate or seal, providing access to the trap

S

safe tray a watertight tray fitted under a feed tank or sanitary appliance to intercept condensation, spillage or leakage, and provided with a waste pipe to direct any discharge to a safe location

sanitation the term used to describe the activities of washing and/or excretion carried out in a manner or condition so that the effect on health is minimised

sanitary plumbing installation an assembly of pipes, fittings, fixtures and appliances used to convey wastes to the sewerage system

scribe to mark by scratching or drawing with a sharp pointed tool on metal workpieces

sedimentation a process in which water is allowed to stand so that the floc and larger particles can settle to the bottom

self-tapping screw a specially hardened screw that taps its own thread in sheet metal and steel

sensible heat that heat which may be felt and measured

separate pipe system an installation in which separate pipes and vents are provided for soil and waste fixtures and in which all waste pipes are connected to the drain through a disconnector trap. Sometimes referred to as a 'double pipe' or 'two pipe' system

sewage 1. *fresh sewage* sewage still containing dissolved oxygen

2. *raw sewage* untreated sewage

3. *septic sewage* sewage in which anaerobic bacteria are present

4. *stale sewage* sewage which has reached an anaerobic (oxygen free) state

sewer the conduit or piping which transports sewage or trade waste, including all accessories

sewerage system the whole system of sewage disposal, excluding the individual property installations

silicon sealant a fluid, resin or elastomer, most frequently used in a jointing compound used in jointing zincalume since it cannot be soldered

silver-brazed joint a welded joint in which the parts are joined with a filler metal which has silver as one of its components

silver brazing (silver soldering) a low temperature brazing process in which a silver alloy is used as a filler metal

single stack system a system of plumbing in which the stack and discharge pipes serve as vent pipes

single stack system (modified) a system of plumbing which differs from the single stack system in that a relief vent is employed and a system of cross-vents are connected to it. This enables a stack of smaller diameter to be used

siphon a pipe system comprising a rising leg and a falling leg, typically in the shape of an inverted U

1. *siphonage* the action of a siphon at atmospheric pressure

2. *induced siphonage* the extraction of water from a trap by siphonage set up by a reduction of pressure at the outlet of the trap

slip joint a joint utilised in guttering to allow fixing of long lengths in situ

socket the 'female' end of a pipe having a larger internal diameter for the reception of the plain or spigot end of another pipe or pipe fitting

soffit 1. (drainage) the highest point of the internal surface of a pipe at any cross-section

2. (roof plumbing) the underside of the overhanging eaves

soil fixture a WC pan, urinal, slop-hopper, autopsy table, bed pan steriliser or sanitary napkin disposal unit

soil pipe the pipe which conveys the discharge from soil fixtures

solvent-welded jointing a method of joining PVC-U materials in which the surfaces to be joined are coated with a solvent which softens or melts the surfaces enabling them to fuse together when brought into contact with each other

specific heat the ratio of the amount of heat required to raise a given mass of a substance through a given temperature range to that required to raise the same mass of water through the same temperature range

spigot the 'male' untreated plain end of a pipe or fitting

spill level the maximum level to which water will rise while overflowing a fixture or storage tank when all outlets are closed, and when the water supply system is discharging at the maximum rate at the nominated pressure

stack a vertical soil or waste pipe extending more than one storey (e.g. 2.5 m) in height

stack vent the extension of a discharge stack above the highest connected discharge pipe. *See also* head vent

stalactite an icicle-like pendant of iron oxide suspended inside a pipe

sterilisation the process of disinfecting water to remove harmful bacteria

stop end the turned-up end of a gutter which prevents water from flowing out

stratification the layering of a body of fluid into two or more horizontal zones according to differing characteristics, especially density

s-trap a trap in which the outlet leg is parallel to the inlet leg

structural slab a self-supporting, concrete floor

sullage a general term for domestic waste liquid other than from soil fixtures, i.e. from bathrooms, laundries and kitchens

sullage pump a small centrifugal pump used for raising sullage from a lower to a higher level

sullage tank a tank used to pre-treat sullage, prior to discharge to a common effluent drainage system

swarf the material removed during cutting and drilling

sweep junction a short length of pipe with a socketed branch at approximately 45°, and with a socket set at 90° to the pipe centre-line

T

temporary bench mark *see* bench mark

test the approved test of pipelines for soundness, as determined by the water authority

test opening an adequate size of opening in a pipe to enable entry of a plug for the purposes of testing

thermal shut down the ability of a thermostatic mixing device to shut down the hot water supply if the device malfunctions

thermoplastic a plastic material that will repeatedly soften when heated and harden when cooled

thermoset a plastic material which is hardened irreversibly by the action of heat and/or a catalyst

thermosiphon the circulation that occurs when gases and liquids are heated; this circulation is modified by varying densities within the substance

throat the inside of a bend

thrust blocks blocks, usually of concrete, placed at intervals along a pipeline and in other positions, adjacent to valves and changes of direction or grade to anchor the pipeline

tinning (tin coating) the process of placing a protective coating over a brass or gunmetal fitting. The coating used is usually an alloy of lead and tin (solder). The term is also applied in brazing and braze welding where the spreading out of a thin layer of fluxed brazing metal ahead of the main deposit to form a 'tinning coat' provides a strong bond between parent metal and deposit

top plate in a timber-framed cottage, the member which takes the main weight of the roof

trade waste waterborne wastes from business, trade or manufacturing premises other than domestic sewage and as defined by the appropriate authority

transpiration a process of transpiring or passing to the atmosphere

trap any fitting designed to retain a water-seal for the purpose of preventing the passage of gases. Traps may also be designed to intercept prohibited discharges

trap vent a vent pipe venting an individual trap to atmosphere or to a main or branch vent pipe to prevent loss of water-seal

trianguiation a method of determining true dimensions by using triangles extending from known fixed points

truss a prefabricated roofing frame. Used as a quick method of pitching roofs

tubing (welding hose) the means by which the gases are supplied from the source (the cylinder) to a welding or cutting blowpipe. Made of reinforced rubber hose, it is strongly built to resist the pressure of the gases and to withstand constant bending and twisting

tundish a small box with a union connection to aerially disconnect a water line from an appliance

U

unvented pipe a discharge pipe without a vent at its upper end

uPVC *See* PVC-U

upstream towards a higher level

upstream vent the vent installed adjacent to the upper end of the drain

V

vacuum breaker device a device designed to introduce air into the system to break the siphon cycle

valve (general) a device used to control the flow of water in a piping system

    1. *service control valve* a valve attached to the main which controls the flow of water along the service pipe to a property

    2. *meter control valve* a valve attached to the service pipe controlling the flow of water through the meter

variable speed a range of speed from one control

vent a pipe provided to limit the pressure fluctuations within a discharge pipe system

vented refers to a discharge pipe having its upper end open to the atmosphere

vent cowl a sanitary fitting for installation on the outlet end of a vent pipe to prevent the ingress of birds, rodents and foreign matter

vertical any pipe which is at an angle equal to or more than 45° to the horizontal. A pipe with a grade of not less than 1:1

W

wail plate a timber strip placed on top of a load-bearing brick wall on which rest the common rafters

waste fixture any waste fixture other than a soil fixture

waste pipe a pipe which conveys only the discharge from waste fixtures

waste water sullage

water-seal the water retained in a trap which acts as a barrier to the passage of air and gas through the trap

wafer table the natural level that underground water maintains

WC the abbreviation for 'water closet'

work lead conductor electrical lead between the source of current and the work or work table

Symbols and abbreviations for sewerage and sanitary plumbing

Element	Symbol
Sewer line	S — Green
Industrial sewerage	—I— IS —I— Green
Soil pipeline	SP Blue
Waste water pipeline	W Yellow
Vent pipeline	V&VP Red
Acid or chemical waste	—\— AW —\— Green
Vent (soil vent) pipe	V (S V)
Agricultural pipe drain	APD
Manhole (all types)	M H
Manhole (all types)	M H
Inspection pit (nature of pit, e.g. dilution, neutralising, designated below symbol)	
Boundary trap	
Inspection shaft	
Grease Interceptor	G
Yard gully with tap	Y G X
Dry pit	
P trap	P
Reflux valve	R
Cleaning eye	
Vertical pipe	Vert
Waste stack	WS
Septic tank	ST
Lamphole	L H
Pumping station	PS
Floor waste	FW
Ejector or pump unit	E

Element	Abbreviation
Down cast cowl	D C
Induct pipe	I P
Mica flap	M F
Junction for future use	J N
Tubs	T
Kitchen sink	K
Water closet	W
Educt vent	E V
Bidet	B I D
Bath waste	B
Handbasin	H
Shower	S
Washing machine	M
Vitreous clay pipe	V C P
Wrought iron pipe	W I P
Cast iron pipe	C I P
Floor waste	F
Urinal	UR
Drinking fountain	D F
Wash trough	W T
Dishwasher	D W
Glasswasher	G W
Bar sink	B S
Cleaners sink	C S
Laboratory sink	L S
Slop sink or slop hopper	S S
Potato peeler	P P
Disposal unit	D U
Cuspidor	CUS
Steriliser	St
Autopsy table	A T
Water meter	W M R
Stainless steel (Corrosion resistant steel)	s s CRS
Hot water unit	H W

Symbols used in plumbing and drainage

Element	Plan	Elevation
Gully (disconnector gully)	100 gully	
Overflow	Same as for above except the use of the abbreviation ORG	
Sealed gully	100 sealed gully	
Inspection opening	IO	
Raised inspection opening or inspection opening to surface	RIO or IOS	100
Floor trap	50 FT	
Floor waste gully	65 FWG	drain
Branch entry to drain	100	
Changing pipe size	65	100	
Changing pipe material	////////////// or use notation	uPVC
Waste pipes	50 T & W 50 FT	50 T & W
100 mm drains and soil pipes	100	
Vents	80 V	50 V or	80 V
S trap pan	WC	
P trap pan	WC	
Soil stack Waste stack	⊙ 100 SS ⊙ 50 WS	100	50

Symbols and abbreviations for water supply and stormwater drainage

Element	Symbol
Water supply	
Watermain (pipeline)	
Domestic water service Cold water	CW
Domestic water service Hot water	HWS
Fire water	FWS Red
Fire hydrant standpipe with cradle and direction of millcock	FH
Spring bail hydrant	SH (SBH)
Fire hose reel	HR (FHR)
Stop valve	S V
Stop tap	S T
Reflux valve	R V
Hose tap standpipe	H T
Water meter	W M
Crossover	
Flanged joint	
Socket and spigot joint	
End capped off	
End blank flanged	
End plugged off	
Flow meter orifice	
Taper	
Cast iron cement-lined pipe	C I C L P
Galvanised mild steel cement-lined pipe	G M S C L P
Asbestos pipe	A C P
Copper pipe	C P

Element	Symbol
Stormwater drainage	
Stormwater line	
Downpipe	
Inspection pit	IP
Inlet gully	IG
Inlet sump	I S
Double grated gully pit	DGGP
Single grated gully pit	SGGP
Grated drain	G D
Open lined drain	OLD
Open unlined drain	OUD
Reinforced concrete pipe	R C P
Vitrified clay pipe	V C P
Glass-reinforced plastic pipe	GRP

Symbolic representation for pipeline drawings

Element	Symbol
Globe valve	
Globe valve with maximum flow adjustment	
Ball valve (spherical plug cock)	
Hand operated valve	
Solenoid operated valve	
Diaphragm operated valve	
Lock shield valve	
Plug cock (tapered)	
Thermostat	T
Humidistat	H
Thermometer dial	T
Pressure gauge with pet cock	P
Sight glass	
Centrifugal pump solid casing	
Centrifugal pump split casing	
Dirt leg with spherical plug cock	
Injector or ejector	
Coil (heating or cooling)	
Bib cock	
Bib cock with hose tail	
Pipe crossing	
Vertical pipe	S
Rising vertical pipe	SR
Dropping vertical pipe	SD
Direction of flow	FLOW
Rise in direction of flow	RISE
Fall in direction of flow	FALL

Element	Symbol
Pipe size (nominated)	20 mm
Reduction in size with branch	50 mm 25 mm / 25 mm
Reduction in size without branch	50 mm 25 mm
Pipe bend	
Pipe elbow	
Pipe tee	
Flanged connection	
Blank flange	
Union joint	
Anchor	A
Hanger	H
Orifice plate	OP
Open vent	
Anti-convection loop	
Pipe guide	
Air cock	AC
Automatic air cock	AAC
Air vessel	
Drain (plug cock with hose tail)	
Thermometer pocket	T
Tundish	
Expansion bellows	
Expansion bend	
Expansion bend (lyre bend)	
Strainer (Y type)	
Strainer (line type)	

Symbolic representation for pipeline drawings

Element	Arrangement drawings		Diagrams A key to the symbols to be provided
	Normal	Skeleton	
Pipe	Large / Small		
Flanged joint			Not required
Spigot and socket joint			Not required
Pipe hanger			Not required
Pipe anchor	Outline of anchor to be drawn	Anchor	Not required
Stop valve			
Stop valve (motor operated)			Ⓜ
Stop valve (right-angle type)			
Stop valve (three-way)			
Spring loaded safety or relief valve			
Counter-weight lever safety or relief valve			
Counter-weight lever safety or relief valve (right-angle type)			
Float operated valve			
Float operated valve (right-angle type)			

Element	Arrangement drawings		Diagrams A key to the symbols to be provided
	Normal	Skeleton	
Non-return or check valve			
Non-return or check valve with lock			
Reducing valve			
Butterfly valve			
Fire hydrant	Outline of hydrant to be drawn	Simplified outline of hydrant to be drawn	
Strainer	Outline of strainer to be drawn	Simplified outline of strainer to be drawn	
Suction pipe strainer			
Suction pipe strainer with foot valve			
Steam trap	Outline of trap to be drawn	Simplified outline of trap to be drawn	
Flow meter (non-recording)			
Flow meter (recording)	R	R	R
Flow meter orifice		Flow meter orifice	
Exhaust head on ventilator			
Pressure or vacuum gauge	Outline of gauge to be drawn	Simplified outline of gauge to be drawn	
Thermometer	Outline of thermometer to be drawn	Simplified outline of thermometer to be drawn	
Steam separator	Outline of separator to be drawn	Simplified outline of separator to be drawn	Alternatives

Conduit diameters (coded)	
20 mm	
25 mm	
32 mm	
40 mm	
50 mm	
100 mm	

Symbolic representation and abbreviations for pipeline drawings

Element	Abbrev.
Air	A
Brine	B
Boiler blowdown	BD
Condensate	C
Cold water	CW
Condenser cooling water	CCW
Oil cooling water	OCW
Engine cooling water	ECW
Chilled water	ChW
Chilled drinking water	Ch DW
Distilled water	DW
Drain or overflow	D
Fuel oil	FO
Feed water	F W
Town gas	TG
Liquefied petroleum gas	LPG
Hot water supply	HWS
High temperature hot water	HTHW
Medium temperature hot water	MTHW
Low temperature hot water	LTHW
Lubricating oil	LO
Oil	O
Nitrogen	N
Oxygen	OX
Nitrous oxide	NO
Refrigerant (Show number if any)	R
Steam	S
Superheated steam	SHS
Vent	V
Vacuum	Va
Demineralised water	DMW
Heat transfer oil	HTO
Pneumatic tube	P T
Fire water service	F WS
Fire foam	F F
Fire sprinkler	F S
Fire CO_2	F CO_2

Element	Symbol
Main piping visible	———— S ————
Main piping concealed	— — — — S — — — —
Piping by others visible	—— - — S —— - —
Piping by others concealed	- - — — - S - — — - -
Future piping visible	— - - —— S —— - - —
Future piping concealed	— — — — S —— — —
Existing piping	———— - S - ————
Piping to be removed	//////////// S ////////////

Index

Exercises

chapter one

- Read the following questions and answer them in the spaces provided.
- You will need to show all your workings where required.
- You can use drawings to help explain your answers.

1   List four available sources of water supply and state their suitability for human consumption.

1 _____

2 _____

3 _____

4 _____

2   Describe the hydrological cycle.

_____

_____

_____

_____

_____

_____

3   What is evaporation?

_____

_____

_____

_____

4   Name the three natural states in which water occurs.

1 _____

2 _____

3 _____

5   What is 'hydraulic gradient' and how does it affect water supply?

_____

_____

_____

_____

_____

6   Explain the importance of gravity in water system plumbing.

_____

_____

7   Explain what is meant by the 'relative density' of water.

_____

_____

_____

8   The minimum height of water in a service reservoir is 1.5 m. Calculate the available water pressure at a fire hydrant which is situated 50 m below the base of the reservoir.

_____

_____

_____

9   The available water pressure in a main is 650 kPa. Calculate the maximum height to which this water may be raised.

_____

_____

_____

_____

chapter two

- Read the following questions and answer them in the spaces provided.
- You will need to show all your workings where required.
- You can use drawings to help explain your answers.

1   Describe the most common method of collecting water for urban water supply.

2   What is a catchment area?

3   What are the four most common functions of a service reservoir?

   1

   2

   3

   4

4   If a service reservoir is 138 metres above the area it serves, what is the static pressure?

5   Elevation (height) is one way of providing pressure. State one other method.

6   Briefly describe how water authorities ensure that the risk of contamination to the drinking water supply is minimised.

_____

_____

_____

_____

7   What are the two main groups of impurities that are found in water?

1 _____

2 _____

8   Explain how these impurities occur in water.

1 _____

_____

2 _____

_____

9   List four causes of odours in water.

1 _____

2 _____

3 _____

4 _____

10 How does 'hardness' occur in water and what effect does it have?

_____

_____

_____

_____

11 What are the two types of water hardness?

1 _____

2 _____

12 What causes each of these types of hardness?

1

2

13 How is hardness removed for each type?

1

2

14 Define the following terms:

Filtration

Sterilisation

Sedimentation

Coagulation

15 Name three alternative sources of water supply for municipal use other than dams.

1 _____

2 _____

3 _____

16 What is a transmission main?

_____

_____

_____

17 What are distribution systems?

_____

_____

_____

18 List two commonly used materials for water mains.

1 _____

2 _____

19 Briefly describe the method of cutting and joining the above materials.

1 _____

_____

2 _____

_____

20 What is the main advantage of laying distribution water mains in a grid pattern?

_____

_____

_____

_____

21 How are metallic mains protected from corrosion?

22 List four advantages of using rubber rings joins in water mains.

1

2

3

4

23 Describe one method of effecting a join in a large diameter pipe/water main.

24 How are metallic mains tapped?

25 How are non-metallic mains tapped?

26 List five factors that affect the classification of a trench to be excavated.

1

2

3

4

5

27 Name and describe three fittings that can be used for repairing water mains.

1 Name _____

Description _____

_____

_____

2 Name _____

Description _____

_____

_____

3 Name _____

Description _____

_____

_____

28 List six instances in which you would use thrust blocks.

1 _____

2 _____

3 _____

4 _____

5 _____

6 _____

29 Sketch (on a separate sheet) one of the instances using thrust blocks listed above.

30 Why are air valves installed in transmission lines?

_____

_____

_____

_____

31 Investigate and prepare a report (on a separate sheet/s) on the source of potable water in your area, its treatment, method of storage and the authority responsible for its supply.

chapter three

- Read the following questions and answer them in the spaces provided.
- You will need to show all your workings where required.
- You can use drawings to help explain your answers.

1   List four different types of piping materials used for water mains.

1 _____

2 _____

3 _____

4 _____

2   Describe the precautions which should be taken to ensure the safety of the public and the plumber when excavating a main for connection purposes.

_____

_____

_____

_____

_____

_____

_____

_____

_____

3   List four methods of installing service pipes under roads.

1 _____

2 _____

3 _____

4 _____

4   Describe the type and position of valves used to control a water supply from main to meter.

_____

_____

_____

_____

_____

_____

5   List two types of basic water meters and describe the method of recording the amount of water in each.

1 Name _____

   Method _____

   _____

   _____

   _____

2 Name _____

   Method _____

   _____

   _____

   _____

6   What method does your local authority employ when connecting services of more than 25 mm diameter to their mains?

_____

_____

_____

_____

_____

_____

7   Investigate and prepare a report (on a separate sheets) on approved materials, other than those covered in this text, for water services in your area. Include in your report any specific requirements for these materials.

chapter four

- **Read the following questions and answer them in the spaces provided.**
- **You will need to show all your workings where required.**
- **You can use drawings to help explain your answers.**

1   List the materials most commonly used in hot and cold water services.

2   List the approved methods of joining any three of the materials mentioned in Question 1.

1 Material

Jointing methods

2 Material

Jointing methods

3 Material

Jointing methods

3   Describe how to make a manipulative joint (flared) when connecting two copper pipes.

4   What are the main advantages of solvent cement joints on PVC–U?

_____

_____

_____

_____

_____

_____

5   Explain the advantages of using cross-linked high density polyethylene (PE–X).

_____

_____

_____

_____

_____

_____

_____

_____

_____

_____

_____

6   List three methods of joining polyethylene (PE) pipe.

1 _____

_____

2 _____

_____

3 _____

_____

chapter five

- Read the following questions and answer them in the spaces provided.
- You will need to show all your workings where required.
- You can use drawings to help explain your answers.

1   List seven control valves or taps.

   1 _____

   2 _____

   3 _____

   4 _____

   5 _____

   6 _____

   7 _____

2   Why is it important that the direction of flow be observed when installing a loose valve stop valve?

   _____

   _____

   _____

   _____

   _____

   _____

   _____

3   Describe the operation of a gate valve.

   _____

   _____

   _____

   _____

   _____

   _____

   _____

   _____

   _____

4   Describe the operation of a float valve.

_____

_____

_____

_____

_____

_____

_____

_____

_____

5   List three pressure control valves and describe the operation of each.

1 Name _____

Operation _____

_____

_____

_____

2 Name _____

Operation _____

_____

_____

_____

3 Name _____

Operation _____

_____

_____

_____

6   Describe the operation of a solenoid valve and state where the valve is most commonly used.

7   Describe the operation of a thermostatic mixing valve.

- Read the following questions and answer them in the spaces provided.
- You will need to show all your workings where required.
- You can use drawings to help explain your answers.

1   Give two reasons why it is necessary to insulate hot water pipes.

1

2

2   List five materials suitable for insulating hot water piping.

1

2

3

4

5

3   What is 'water hammer' and how does it affect a water service?

4   List and describe the three main factors that influence the intensity of water hammer.

1   _____

_____

_____

2   _____

_____

_____

3   _____

_____

_____

_____

5   List five methods for reducing or eliminating water hammer in pipes.

1   _____

_____

2   _____

_____

3   _____

_____

4   _____

_____

5   _____

_____

- **Read the following questions and answer them in the spaces provided.**
- **You will need to show all your workings where required.**
- **You can use drawings to help explain your answers.**

1 Define the following terms:

Heat

Sensible heat

Latent heat

Specific heat

2 What units of measurement are used for heat energy?

3 What is the specific heat value of water when measured in MJ?

4 What is the co-efficient of expansion for water?

5 Calculate the energy required to raise the temperature of 400 litres of water from 15 °C to 75 °C using the following formula:

HE = L x T x C

Explain the formula:

where HE = _____

L = _____

T = _____

C = _____

Working: _____

_____

_____

_____

6 If the above 400 litre water heater were fuelled by natural gas and rated at 45 MJ/hr, how long would it take to raise the temperature of the water from 15 °C to 75 °C?

_____

_____

7 How much expansion would occur in the above heating cycle? Use the following formula

E = L x T x C

Explain the formula:.

where E = _____

L = _____

T = _____

C = _____

Working: _____

_____

_____

_____

8 Define the following terms:

Conduction ....

....

....

Convection ....

....

....

Radiation ....

....

....

9 With the aid of a sketch illustrate how heat is transferred by conduction, convection and radiation.

10 What are the four most common sources of energy for heating water?

1 _____

2 _____

3 _____

4 _____

11 The energy efficiency rating of gas storage water heaters has been greatly improved in recent years. How has this been achieved?

_____

_____

_____

12 What is meant by the term 'off-peak' in relation to electricity charges?

_____

_____

13 Why are heat pumps more energy efficient than other electric systems?

_____

_____

_____

14 Why is auxiliary heating necessary in solar water heaters?

_____

_____

_____

15 What is a 'wetback' and where would it be installed?

_____

_____

_____

16 In order to meet sustainability targets in building design, high efficiency heaters are to be used. What type of water heaters would you suggest?

_____

_____

17 Hot water systems may be divided into two basic types. What are these types?

1 _____

2 _____

18 List the water heaters for each of the types in Question 17.

19 For each of the types of water heaters listed in Question 17, choose two from each type and on a separate sheet describe the advantages and disadvantages of each and their installation requirements. Include a drawing of the installation connections, including all valves and ancillaries, and manufacturer's brochures.

20 Describe the refrigeration cycle used in a heat pump, including how it provides energy to heat the water.

21 Describe the thermosiphon principle.

22 Sketch a boiler/cylinder system.

23 What are the two methods of providing warm water?

1 _____

2 _____

24 How is the build-up of bacteria or viruses controlled in a circulating warm water system?

_____

_____

25 What is the principle behind 'equaflow'?

_____

_____

_____

26 Why is it important to ensure that the equaflow principles are followed?

_____

_____

27 Where the hot and cold water enters from the same side of a multiple water heater installation, what special piping consideration needs to be observed to ensure equaflow principles are maintained and how can we check this?

28 What are mains pressure water heater cylinders commonly constructed of?

29 Sketch the installation set-up of your water heater at home. Does the installation meet AS/NZS 3500 requirements? Include relevant clauses. If not, what needs to be done to make it comply?

30 Describe two common methods of testing hot and cold water systems.

1

2

31 List nine operations which should be completed when commissioning an installation.

1 _____

2 _____

3 _____

4 _____

5 _____

6 _____

7 _____

8 _____

9 _____

32 Explain where and why a test bucket is used in the installation of hot and cold water systems.

_____

_____

_____

_____

_____

_____

_____

_____

_____

_____

33 What is the hydrostatic test pressure and for how long a period is the test carried out?

_____

_____

_____

_____

_____

- **Read the following questions and answer them in the spaces provided.**
- **You will need to show all your workings where required.**
- **You can use drawings to help explain your answers.**

1 List four sources of water supply for domestic use on rural properties that are not served by mains supply.

1

2

3

4

2 List three materials that may be used for the construction of rural water storage tanks.

1

2

3

3 List and describe the two types of groundwater that are common to rural areas.

1

2

4   Describe the term 'pressure system' with reference to rural water supply and explain where a pressure system can be used.

- Read the following questions and answer them in the spaces provided.
- You will need to show all your workings where required.
- You can use drawings to help explain your answers.

1   Explain the following terms:

static head _____

_____

_____

_____

_____

pressure head _____

_____

_____

_____

_____

friction head _____

_____

_____

_____

_____

2 Describe two situations where plumbers may need to use pumps during construction work.

1 _____

_____

_____

_____

2 _____

_____

_____

_____

3 Why should controls on the delivery side of a pump be kept to a minimum?

_____

_____

_____

_____

_____

_____

_____

_____

_____

_____

4 Describe a situation where a pump may need to be installed for domestic purposes and the type of pump required.

5 Explain the operation of a centrifugal pump and describe a situation where the pump may be used.

6   List and describe the four most common restrictions concerning groundwater supply.

1 _____

_____

2 _____

_____

3 _____

_____

4 _____

_____